Anonymous

**Memoirs of the Geological Survey**

England and Wales. Sheet memoirs (o.s.). A.

Anonymous

**Memoirs of the Geological Survey**
*England and Wales. Sheet memoirs (o.s.). A.*

ISBN/EAN: 9783337322915

Printed in Europe, USA, Canada, Australia, Japan

Cover: Foto ©berggeist007 / pixelio.de

More available books at **www.hansebooks.com**

80 S.W.

# MEMOIRS OF THE GEOLOGICAL SURVEY.

## ENGLAND AND WALES.

THE

# GEOLOGY OF THE NEIGHBOURHOOD OF CHESTER.

(EXPLANATION OF QUARTER SHEET 80 S.W.)

BY

AUBREY STRAHAN, M.A., F.G.S.

PUBLISHED BY ORDER OF THE LORDS COMMISSIONERS OF HER MAJESTY'S TREASURY.

LONDON:
PRINTED FOR HER MAJESTY'S STATIONERY OFFICE,
AND SOLD BY
LONGMANS & Co., Paternoster Row ; TRÜBNER & Co., Ludgate Hill ;
LETTS, SON, & Co., Limited, 33, King William Street ;
EDWARD STANFORD, Junior, 55, Charing Cross ;
J. WYLD, 12, Charing Cross ; and
T. J. DAY, 53, Market Street, Manchester :
ALSO BY
Messrs. JOHNSTON, 16, South St. Andrew Street, Edinburgh :
HODGES, FIGGIS, & Co., 104, Grafton Street, and A. THOM & Co.,
Abbey Street, Dublin.

1882.

Price Two Shillings.

# NOTICE.

THIS Quarter Sheet (80 S.W.) was surveyed by Prof. Hull in 1855. The superficial deposits were surveyed by Mr. Strahan in 1878, when a complete revision of the boundaries of the subdivisions of the Trias was made, the Keuper having been entirely re-mapped on the six-inch scale. A new line has been added, separating the Waterstones from the Conglomeratic Sandstones beneath, while the base line of the Red Marl of the Keuper formation has been almost entirely redrawn. At the same time two faults of great importance have been proved to exist (viz., the Great Delamere Fault, and the North and South Red Marl Boundary Fault, at Delamere), besides which numerous other smaller dislocations have also been mapped, while the positions of those shown in the first edition have been corrected.

A List of Works on the Geology of Cheshire has been added to the Memoir by Mr. Whitaker.

HENRY W. BRISTOW,
Senior Director.

Geological Survey Office,
28, Jermyn Street,
London, S.W.

# CONTENTS.

# LIST OF WOODCUTS.

# GEOLOGY

OF THE

# NEIGHBOURHOOD OF CHESTER.

## PART I.

## TRIASSIC ROCKS.

### INTRODUCTION.

THIS Quarter-sheet includes an area of 168 square miles, extending from Chester on the west side to Calveley, Little Budworth, and Acton on the east; and from Helsby and Kingsley on the north to Burwardsley and Bunbury in the south. This district includes a portion of the course of the Dee, but is for the greater part drained by the Gowy and the Weaver, tributaries of the Mersey. The Watershed between the Dee and the Mersey drainage systems runs from Whitby between Caughall and Wervin, by Christleton and Waverton, north of Tattenhall, to near Burwardsley.

The western portion of this area is underlain by the three sub-divisions of the Bunter Sandstone; the eastern includes a part of the great tract of Keuper Marl, which extends to Macclesfield on the east, and to near Shrewsbury on the south. The areas occupied by the Bunter Sandstone and the Keuper Marl are separated by a line of hills and inland cliffs ranging north and south. These hills mark the outcrop of the Waterstones and the hard Building-stones and Breccias of the Basement Beds of the Keuper. The whole of these rocks belong to the Triassic group.

The succession of beds, in descending order, is as follows :—

|  |  |  |  |
|---|---|---|---|
| KEUPER | - | - | - ⎰ Keuper Marl.<br>⎱ Waterstones.<br>Basement Beds. |
| BUNTER | - | - | - ⎰ Upper Mottled Sandstone.<br>⎱ Pebble Beds.<br>Lower Mottled Sandstone. |

### BUNTER.

Owing to the prevalence of Drift little is seen of the rock in the Bunter area, except in hills marking the outcrop of the hard conglomeratic sandstone of the Pebble Beds. These hills form two very distinct ranges, the first running through Tattenhall,

Tarvin, Barrow, Dunham, and Hapsford; the second through Aldford, Eccleston, Chester, and Whitby; a third less distinct passes through Saighton and Christleton. The dip of the rocks in these hills is uniformly to the east, and the line of strike is indicated by the direction in which they range. The low ground separating the ranges is deeply obscured by Glacial Deposits.

In the absence of further evidence, it might have been inferred that this form of ground indicated the existence of three beds of conglomerate rising one from under the other towards the west; but the three-fold division of the Bunter having been proved by Prof. Hull to exist as given above over the whole of the Triassic areas of central England,* it has been assumed by him that the same bed of conglomerate has been twice thrown down to the west by the agency of large north and south faults, and consequently appears at the surface three times. The intervening low ground is assumed to be occupied by the Upper and Lower Mottled Sandstones respectively. The supposition of the three ranges being due to successive beds of conglomerate would imply a great thickness of Bunter rocks, but if the theory of the faults is correct, the base of the Trias might be found at comparatively slight depths in suitable localities, namely, where the Lower Mottled Sandstone rises to the surface.

*Lower Mottled Sandstone.*—In the area supposed to be occupied by this subdivision, exposures of the rock are rare. Soft red sand forms the surface near Handley Church, and soft red and white sand is exposed in the railway-cutting at Waverton, where it is faulted against Pebble Beds. At Hapsford similar beds are to be viewed in several places, and are seen to pass under the Pebble Beds of Hapsford and Dunham to the east. The Waverton and Handley soft sands appear to have this relation also to the Pebble Beds of Tattenhall and Tarvin. Bright red, yellow, and white sand, occasionally composed of round loose grains of quartz, is exposed in the cutting of the Cheshire Lines near Hoole. On the north they are faulted against Pebble Beds. Soft red sand-stone belonging to this subdivision was formerly exposed to view in the road near the canal south of Little Mollington.

The Lower Mottled Sandstone consists of bright-coloured yellow and red or white soft sandstone, without pebbles, but with oblique bedding, indicating the action of currents. The variations in colour are the result of subsequent alteration; the white patches in the red rock occasionally contain a nucleus of iron pyrites or a trace of lime. The thickness of this subdivision is very variable, but probably does not exceed 500 feet. Should the Permian rocks be absent in this district, the Lower Mottled Sandstone will rest on Carboniferous rocks.

*Pebble Beds.*—The distribution of this rock has been partly described above. It consists of moderately hard fine-grained sandstones of a red colour, with occasional white bands or patches, and with well-rounded pebbles of liver-coloured and white quartz,

_____

* The Triassic and Permian Rocks (Geol. Survey Memoir), p. 29.

scattered through the rock, or forming shingly partings between beds of freestone. Thin incontinuous shales are also interstratified with the sandstones.

The Pebble Beds have been quarried in many places for building purposes, but their inferiority for this purpose is shown by the condition of Chester Cathedral (before restoration), and St. John's Church-tower. A great portion of the stone used in the Cathedral and the old Abbey buildings adjoining it was obtained from quarries between the Northgate Street and Windmill Lane, and in places within the walls on the east side of the Northgate Street.* Most of these have been long since abandoned, and are in part filled up and built over. The same rock is well exposed in the Canal and Railway cuttings. The outcrop of the Pebble Beds on which Chester stands forms the most westerly of the three ranges of hills or escarpments mentioned above. The river Dee after following this escarpment for some miles cuts clean across it at Chester in a narrow steep-sided valley, excavated entirely in the rock. Along this valley the beds are frequently exposed. They rise in the bed of the river below the Grosvenor Bridge, and run along both sides of it as far as the Grosvenor and Queen's Parks. Further east they have been proved in tunnelling under the river at the waterworks at Boughton. At several places the dressed faces of the rock indicate that it has been quarried. I am informed† that an old pit, filled with quarry-spoil, was found in making additions to the gaol, and an excavation resembling a quarry, but below the level of the river, was met with in placing the foundations of the north end of the Grosvenor Bridge. It is probable that many of the quarries in Chester are of great antiquity, as this stone was used by the Romans in portions of their buildings; but that they were aware of its inferiority as a building material is shown by their having used large and well-trimmed blocks of the sandstones of the Keuper Basement Beds for the facing of their city wall. In mediæval times the local stone was more universally worked, and little more than crumbling ruins remain of the buildings constructed of this poor material. Most of the villages of the Bunter area are situated on the hills formed by the Pebble Beds, and in all of these the church and other principle buildings have been constructed of the Pebble Bed Sandstone obtained from quarries in the immediate neighbourhood. The principal quarries are in the villages of Eccleston, Christleton, Waverton, Saighton, Handley, Tattenhall, and Tarvin. The stone seems to last better when not exposed to the vitiated atmosphere of a town.

The thickness of this subdivision has been estimated at 600 feet,‡ but may probably be more. It was found to exceed 1,200 feet in a borehole at Bootle, near Liverpool.

*Upper Mottled Sandstone.*—The highest member of the Bunter consists of bright red, yellow, or white soft sandstone, or red

* The positions of these were pointed out to me by Mr. Shrubsole, F.G.S.
† By Mr. Manning, Governor of Chester Castle.
‡ The Triassic and Permian Rocks (Geol. Survey Memoir), p. 61.

A 2

sandstone with white blotches, without pebbles, and closely resembling the Lower Mottled Sandstone. The lower portions of the subdivision and its junction with the Pebble Beds are nowhere exposed within this area, but there are numerous good sections of its upper part and junction with the overlying Keuper Beds in the Helsby, Delamere, and Peckforton Hills. The following details were taken in descending Helsby Hill by the lane leading from Alvanley to Helsby Station above Helsby Quarry :—

KEUPER SANDSTONE FORMING THE CRAGS.*

|  |  | feet. |
|---|---|---|
| BUNTER | Mottled sandstone, and red fine-grained sandstone with a coarser bed containing fragments of shale - - - | 25 |
|  | Soft red and mottled sandstone, current-bedded in long sweeping planes - - - - - - - | 30 |
|  | Dull red and grey sandstone, full of rolled lumps of shale about as big as a shilling - - - - - | 6 |
|  | (a.) Shale, holding up a little water - - - - | 2 |
|  | Fine-grained, dull red and grey sandstone, current-bedded and with rolled lumps of red and grey shale. In the upper part 2 feet of grit, consisting of well-rounded grains of quartz loosely cemented - - - - - | 45 |
|  | Red shale, not continuous, up to 8 inches - - - | 0 |
|  | (b.) Grey quartzose grit, closely resembling in texture and colour Keuper Sandstone - - - - - | 1 |
|  | Soft yellow and red sandstone - - - - - | 20 |
|  | Grey sandstone, with soft red shaly sandstone; with white bands and mottles; and some harder quartzose sandstones dull in colour - - - - - - - | 60 |
|  | Soft current-bedded sandstone, bright-red with white mottles, elongated in the direction of the current-bedding planes, which form sweeping curves 50 yards in length - - | 50 + |

The letters (a) and (b) refer to the accompanying section Fig. 1. The beds so marked are exceptional in the Upper Mottled Sandstone, and in their conglomeratic nature and quartzose grain resemble parts of the Lower Keuper Sandstone. They are not continuous, and are overlaid by thick beds of the Bunter type, over which occurs the Keuper building stones with a well-defined base. A similar instance occurs in Thurstaston Hill (Explanation 79, N.E.), where a hard quartzose band resembling the Lower Keuper beds occurs in the Upper Mottled Sandstone.

It is noticeable in the preceding section that the tint and vividness of the colour is dependent on the grain of the stone, the coarser-grained quartzose sandstones having the pale tint of the Keuper, while the fine-grained or loamy varieties have the brightness common in the Bunter, irrespective of the age of the rock.

The abrupt variations in colour in these beds is nowhere better seen than at Beeston Castle. The approach to the outer fortifications is cut in soft red sandstone, which continues to the base of the Keuper Sandstone at the old gateway. On the north side of the hill the same beds are yellow with irregular seams of iron oxide, and on the north-west they are chiefly yellow, but contain a few red patches. On the west side the upper 6 feet are yellow and

---

* For details, see page 7.

FIG. 1.

Section from Helsby Marsh through Helsby Hill to Alvanley.

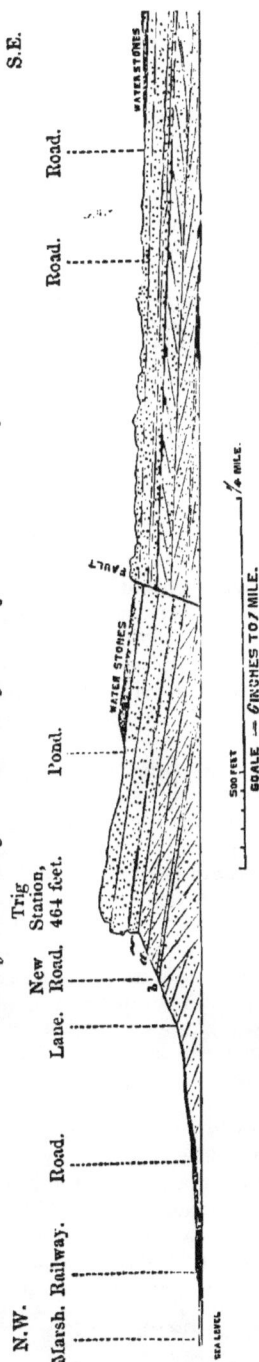

the remainder red, while from the south-west to the south-east angle, the upper part is red and the lower composed of the thick beds of white sand in which some caves have been excavated.

The red colour, which was once probably diffused through the whole mass of rock, is due to a film of peroxide of iron coating the grains of the sandstone, and cementing them into a coherent stone. Those portions of the rock which are now colourless have been bleached in consequence of the solution and removal of this film of peroxide of iron, probably by carbonated water. It may be remarked that some colourless patches of rock are surrounded by more or less concentric seams of hydrated sesquioxide of iron, as though the iron had been acted on and dispelled from within; while in others the absence of any enclosing seam, and the existence of a central nucleus, usually of pyrites, indicates an opposite process, namely the concentration of the iron inwards to form a nodule, and its simultaneous conversion into pyrites by combination with hydro-sulphuric acid, very probably derived from the decay of some organism.*

In the Upper Mottled Sandstone, and in similar fine-grained and loamy beds in the Keuper Basement Beds, the faults, joints, and minute crevices are filled in with a hard siliceous deposit, ringing to the hammer, and standing out in sharp razor-like edges, as the surrounding softer rock crumbles away. This vein-deposit is not found in the more open quartzose sandstones.

This subdivision is estimated to be from 500 to 800 feet in thickness.

## KEUPER.

*The Basement Beds.*—The lowest beds of the Keuper consist of hard brown and white quartzose grits and breccias. The superposition of these beds on the soft Upper Bunter has given

b. Coarse grit-bed in the Upper Mottled Sandstone.
a. Thin shale in the Upper Mottled Sandstone.
m. Junction of Bunter and Keuper.

SCALE — 6 INCHES TO 1 MILE.

500 FEET

¼ MILE.

S.E.

Road.

Road.

Road.

WATER STONES

FAULT

WATER STONES

Trig Station, 464 feet.

New Road.

Lane.

Road.

N.W.

Marsh. Railway.

SEA LEVEL

* See also Maw on Variegated Strata. Quart. Journ. Geol. Soc., vol. xxiv., p. 351.

rise to the series of escarpments by which the great tract of Keuper Marl is almost entirely fringed.  Owing to the prevailing easterly dip, the steep faces of the hills are turned towards the west, their lower slopes being occupied by the Bunter, and the tops crowned by the crags of the Keuper Grits; the gentle slopes descending towards the east are underlain by the Waterstones.  These features are exemplified in Helsby Hill (Fig. 1, p. 5), and are reproduced wherever the beds are repeated by faults.

The sandstones occur in three or more courses separated by partings of "roach" (soft sandstone) or shale.  They have a coarse quartzose grain, and contain a few small pebbles, averaging the size of a bean, scattered throughout.  The beds of roach are fine-grained, bright-coloured, current-bedded, and not conglomeratic, so that they closely resemble parts of the Bunter.  The alternation of the hard sandstones with these softer beds of roach has given rise to the terraces and minor escarpments observable in Helsby Hill and others.

The bases of the sandstones are frequently brecciated and contain pebbles more abundantly than the upper parts.  In such cases the surface of the underlying "roach" on which they rest presents an appearance of erosion.  All the beds of hard sandstone possess these characteristics in common, but from its comparative constancy and the ease with which it could be followed on the ground, the base of the lowest conglomeratic sandstone was traced upon the map.  It was ultimately considered to be the base of the Keuper formation, and the signs of erosion of the underlying bed, the cutting off of the current-bedding planes and the change of colour were considered to be indications of unconformity between the Bunter and Keuper Series.

But it has been shown above that the tint is dependent upon the grain of the rock, the finer and more loamy beds of either the Keuper or Bunter being characterised by bright colouring.  The cutting off of the current-bedding places and the erosion merely indicate a change either in the strength or direction of the currents.  These appearances, moreover, are not only repeated at the base of every conglomeratic course of sandstone in the Keuper, but are also observable at the base of the Pebble Beds of the Bunter.*  Taking into consideration the close similarity of the conglomeratic beds of the two ages, and the repetition of all the phenomena in the one that are observable in the other, the conclusion is inevitable that in this area the deposition of the Keuper followed on that of the Bunter under a continuance of the same physical conditions.

The junction of the Keuper and Bunter is visible in Frodsham Hill (Dunsdale Hollow, 80 N.W.); in a lane running east of Woodhouse Hill; in a small outlier faulted down west of Woodhouse Hill; in the caves at the north-east corner of Helsby Hill, and in the north-west cliff of Helsby Hill at the foot of the crags from near the quarry for 500 yards northwards as far as a small fault which throws down the beds 12 feet to the south-west, and

---

* The Triassic and Permian Rocks (Geol. Survey Memoir), p. 35' 57.

has caused a cleft in the rocks ; north-west of Manley Quarries ; at the south end of the Helsby and Manley outlier, in a wood and in a road-cutting; at the north end of Alvanley Hill ; near Riley Bank ; at the north end of Simmonds Hill; on the north side of Willington Corner ; at Beeston Castle, at the old gateway and along the south-west and west cliffs about 30 feet below the south-west angle of the Castle (Frontispiece).

No more complete section of these beds can be obtained than in Helsby Hill (Fig. 1, p. 5). The fern-covered slopes of the Bunter are broken only by the hard beds in the Upper Mottled Sandstone (*a*) and (*b*). The partings between the building-stones form a shelf in the cliff and a terrace near the top. From the summit eastwards the ground slopes with the dip of the beds to a depression which marks the position of some soft sandstones forming the top of the Basement Beds (Frodsham Beds). These are succeeded naturally by the Waterstones which are exposed in the marl-pit. The details of a descending section in the crags are as follows :—

| | feet. |
|---|---|
| Hard grey sandstone, with few small white quartz pebbles - | 30 + |
| Grass-covered terrace, marking the position of a soft bed - | — ? |
| Hard building-stone, with a few pebbles - - - | 45 |
| Coarse grit, hackley rock and breccia with lumps of shale, fragments of sandstone and quartz-pebbles - - | 5 to 10 |
| Line of large lumps of shale, apparently remnants of a broken-up shale-bed - - - - - | ½ to 0 |
| Hard building-stone with few small quartz-pebbles, the base occasionally brecciated, and containing fragments of sand-stone - - - - - - - | 35 |
| Soft, mottled, red and white roach (Bunter) - - — |  |

The section in the two lower courses of sandstone in the quarry at 650 yards distance westwards is as follows :—

| | feet. |
|---|---|
| Sandstone, good building-stone - - - - | 25 + |
| Soft, white, current-bedded sand with two 8-inch bands of dark-red shale - - - - - - | 5 |
| Do.    do.  with irregular red seams - - - | 10 |
| Dark-red shale with bullions - - - - | 7 |
| Sandstone, good building-stone - - - - | 40 + |

A descending section at Beeston Castle is as follows :—

| | feet. |
|---|---|
| Hard sandstone with a few quartzite pebbles and fragments of soft sandstone, which by crumbling out have left cavities - - - - - - - | 10 |
| Breccia, not continuous - - - - - | 0 to 4 |
| Soft, mottled roach, like Bunter - - - - | 3 |
| Hard sandstone with cavities - - - - | 18 |
| Soft sand-parting - - - - - - | 1 |
| Hard, pale-red hackley sandstone, brecciated at the base with fragments of soft sandstone, occasionally hardened with a siliceous cement - - - - - | 16 |
| Soft, current-bedded roach (Bunter) |  |

The joints in the Keuper Sandstone at Beeston Castle contain abundance of Heavy Spar (Sulphate of Barium), those at Helsby have also a trace. This mineral, with carbonate of lime, is very abundant in the breccias of the Peckforton Hills. A structure of doubtful origin* occurs also in the stone, giving it a nodular appearance. It is visible only on surfaces that have been exposed to weather or subjected to friction, such as those of doorsteps, where it gives rise to small oblong prominences, about half an inch long, and one tenth broad. Occasionally two are found to intersect so as to produce a cross. The grains of sand in these prominences are similar to those in the surrounding stone, but are cemented by a whitish mineral, which, in crystallising out, seems to have bound them together into an obscure crystalline form. This mineral gives no effervescence with acids, but shows a sulphur reaction under the blow-pipe. It is not Sulphate of Lime, but is probably Sulphate of Barium.

The building-stone obtained from these beds is superior to that of the Pebble Beds of the Bunter, and has been used to replace it in the restoration of Chester Cathedral. Its usual colour is a subdued brownish-red, occasionally mottled with white; more rarely it is pure white. The chief quarries are at Helsby, Manley, Simmond's Hill, Delamere, Kelsall, and Peckforton.

The quarry at Delamere is traversed by strong joints running S. 25° E. parallel to a large fault which runs at the foot of the slope and brings in Keuper Marls under the Forest. The following is the descending section :—

|  | feet. |
|---|---|
| Worthless stone, full of pebbles and soft fragments | 0 to 4 |
| Brown sandstone, with a few small quartzite pebbles | 16 |
| Coarse, loose grit with quartzite pebbles and very numerous cavities containing a loose brown earth and fragments of shale, current-bedded to N.W. | 0 to 1 |
| Brown sandstone | 18 + |

South of this quarry and near Delamere Rectory a thick series of slates and flaggy, micaceous sandstones appear in the Basement Beds. They somewhat resemble the overlying subdivision or Waterstones, but their true position is proved by their being overlain by a conglomeratic sandstone with pebbles of quartzite. A descending section shows :—

|  | feet. |
|---|---|
| Brown sandstone, with quartzite pebbles | 15 + |
| Soft, red, even-bedded sandstone | 4 |
| Shale and sandy marl, with beds of sandstone wedging in | 8 |
| Soft shaly sandstone, with shale partings | 7 |
| Red shaly marl, with beds of fine-grained white sandstone wedging in | 6 + |

The beds immediately underlying these were passed through in a well close by and found to consist of brown grits and white micaceous flags of the usual Basement Bed type. The shales thin out rapidly in every direction.

Beds of shale in the Basement Beds of 3 to 4 feet thickness occur also at Woodhouse Hill; near the Fish-pool; at Willington Corner; near Alvanley; and near Helsby and Frodsham. They are characterised by inconstancy.

---

* This was first pointed out to me by Mr. John Price of Chester.

In a quarry near Mouldsworth Station, Lower Keuper Sand-stone is seen to be faulted against Upper Mottled. The former is very much shattered and white. The fault, which is about 12 feet wide, is filled with the débris of the two rocks. One mile north-west of this point occurs the pure white building-stone of Manley. This stone is continuous with that of Helsby, and is brought up to the surface at Manley by a reversal of the dip at Clough Moors, in consequence of which the beds form a basin. The Manley Stone occurs in a bed about 45 feet thick, the best quality being at the bottom. The grain is coarse and sharp, and generally too loose to make a first-class building material. Rolled lumps of green shale and small pebbles of quartzite occur in it. It is overlain by soft and current-bedded white sands and red shaly sandstone (Frodsham Beds); southward several courses of a similar stone crop from beneath it, showing that it occupies a position near the top of the Basement Beds. The stone has been used in the Grosvenor Bridge and Chester Castle, including the monoliths supporting the portico, and in parts of Eaton Hall.

The upper portion of the Basement Beds consists of soft, current-bedded and bright-red, white, and yellow sandstones, to which the name of *Frodsham Beds* has been given in consequence of their fine exposure in the railway cutting at that place (Explanation 80 N.W.). They so closely resemble the softer subdivisions of the Bunter as to be distinguishable only by their relation to the conglomerates below and the Waterstones above. They are constant over the present area, but very variable in thickness, owing to the wedging in of beds of conglomerate at various horizons in them. Exposures occur in the lanes west of Newton, near Frodsham; on Eddisbury Hill; below the trench of the old camp; and in Organ's Dale: in all which places they are of a bright red colour. At Manley, south of the Heald, and in Kelsall they are principally yellow; but to the south-east round Tirley Farm and High Billinge they are again bright red. They are finely exposed in the Holbitch Slack, and near the Yew Tree House, but south of this point are cut out by faults which throw the Waterstones against the lower part of the Basement Beds, or against the Bunter series.

The tracks of *Cheirotherium* have been noticed in the Basement Beds of Daresbury,[*] Storeton, Weston,[†] and Frodsham.[‡] This subdivision varies from 180 to 250 feet in thickness.

*The Waterstones.*—The succeeding subdivision received this name§ from the abundance of water contained in it in consequence of the alternation of beds of impervious shale with porous sand-stones. It contrasts strongly with the underlying subdivisions in the perfect regularity of the bedding, the loamy texture of the sandstones, and the absence of the conglomeratic character. The line of separation between the Waterstones and the Frodsham

* Williamson, Quart. Journ. Geol. Soc., vol. xxiii., p. 56, 1867.
† Black, Quart. Journ. Geol. Soc., vol. ii., p. 65. 1846.
‡ Geology of the Country around Prescot, 3rd edition. (Now in press.)
§ From Messrs. Binney and Ormerod. See Discussion on a paper by Prof. Hull, Trans. Geol. Soc. Manchester, vol. ii., p. 32.

Beds is moreover sharp, each formation preserving its distinctive character up to the actual contact. This line, which can be followed with great ease owing to this circumstance, has been drawn upon the Quarter-sheet now for the first time.

The junction is visible at the south-east corner of Eddisbury Camp; in Kelsall Village; south-east corner of the Heald; in Delamere Forest near the fault; in a lane 300 yards west of the Forest; at a junction of lanes 500 yards north of High Billinge; at Quarry Bank about half-mile south of this point; and in Holbitch Slack. The accompanying sketch was taken at the last-named locality.

FIG. 2.

*Junction of the Waterstones and Frodsham Beds.  Holbitch Slack.*

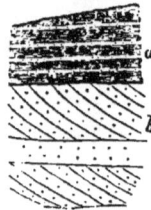

a. Waterstones.            b. Frodsham Beds.

The current bedding planes of the bright red Frodsham Sandstones are cut off abruptly by nearly horizontal shales and flags forming the base of the Waterstones. The section is similar to those observed near Frodsham and Runcorn (Explanation 80 N.W.). A little higher up the ravine a small north and south fault throws down the junction below the level of the brook.

The certainty with which this boundary can be identified has rendered it possible to trace many important lines of fault. These will be more fully described hereafter, but it may be mentioned that isolated patches of Waterstones have been preserved by their agency in the area occupied by the Basement Beds, as at Kelsall. In other cases, as at Eddisbury Hill, High Billinge, and Tirley Farm, the Waterstones cap hills whose flanks are formed by the Frodsham Beds and Conglomerates. South of Utkinton the whole range is formed by Waterstones, the Basement Beds being cut out by a large fault. But at Beeston Castle and Peckforton, where the Waterstones have been completely denuded away, the Basement Beds present their usual picturesque characteristics of crag and cliff.

Sections in the Waterstones are of frequent occurrence in the pits opened to obtain marl for spreading on the land. In its lower part the subdivision consists of red and green shales with thin flags and thin loamy sandstones, as exposed at Helsby, Finney Hill, Eddisbury, Kelsall, High Billinge, and Holbitch Slack.

A descending section on Helsby Hill is as follows:—

|                                              | ft. | ins. |
|----------------------------------------------|-----|------|
| Shattered flaggy sandstone, brightish red -  | 3   | 0 +  |
| Red marl -                                   | 1   | 0    |
| Green shale                                  | 1   | 0    |

| | | | | | | ft. | ins. |
|---|---|---|---|---|---|---|---|
| Red, sandy marl with grey bullions | - | - | - | | - | 1 | 6 |
| Stony red band | - | - | - | - | - | 0 | 8 |
| Red marl - | - | - | - | - | - | 0 | 10 |
| Shale with stony bands, pale green | | | - | - | - | 0 | 10 |
| ,, ,, pale red | | | - | - | - | 0 | 8 |
| ,, ,, pale green | | | - | - | - | 0 | 6 |
| ,, ,, pale red | | | - | - | - | 7 | 0 |
| Red shaly marl - | - | - | - | - | - | 1 | 0 |
| Green shale | - | - | - | - | - | 2 | 0 |
| Pale red, stony shale | - | - | - | - | - | 8 | 0 + |
| Not seen; about | - | - | - | - | - | 10 | 0 |
| Frodsham Beds - | - | - | - | - | - | — | |

Finney Hill Marl-pit, descending :—

| | | | | | | ft. | ins. |
|---|---|---|---|---|---|---|---|
| Red shaly marl | - | - | - | - | - | 20 | 0 |
| Sandstone | - | - | - | - | - | 4 | 0 |
| Shale and sandy beds | - | - | - | - | - | 15 | 0 |
| Sandstone - | - | - | - | - | - | 6 | 0 |
| Green and grey sandy marl | - | - | - | - | 9 | 0 |
| Red sandy marl - | - | - | - | - | - | 10 | 0 |
| Green shale ⎤ | | | - | - | - | 1 | 0 |
| Red marl - ⎬ good for marling land | | | - | - | - | 8 | 0 |
| Green shale ⎦ | | | - | - | - | 0 | 9 |
| Red sandy marl | - | - | - | - | - | 3 | 0 + |

At Irk Wood, near Mouldsworth, there are red and green shales (of good quality for marling purposes), faulted against grey and red sandstone with shales. They have been used for marling Simmond's Hill.

In the Camp on Eddisbury Hill the base of the Waterstones consists of even-bedded red marl with a few thin flaggy bands.

At Kelsall the same beds are white, and consist of thin-bedded flags and shales. At a higher level in the quarry on Longley Hill these occur in descending order :—

| | | | | ft. | ins. |
|---|---|---|---|---|---|
| Soft yellow and red flaggy sandstone | - | - | - | 11 | 0 |
| Reddish shale | - | - | - | - | 3 | 0 |
| Sandstone | - | - | - | - | 3 | 0 |
| Red shale; green at the bottom | - | - | - | 3 | 6 |
| Sandstone - | - | - | - | - | 3 | 0 |
| White, micaceous, flaggy sandstone (good) - | - | 7 | 0 |

The stone is of good quality; it splits readily into slabs, and is worked into hearth-stones, window-sills, &c. The surfaces of the shales show beautifully preserved ripple-marks.

The upper portion of the subdivision consists of thick even-bedded sandstones, fine-grained and not conglomeratic, interstratified with continuous laminated shales. They are exposed to view in Delamere Forest north and east of Kingswood; at Luddington Hill; and in numerous places near Eaton and Tarporley. They everywhere throw out water, but in unusual abundance in some springs along a probable line of fault running from near Kingswood to Wilkinson's Lodge. Several springs issue from the Waterstones in the hill-side above Eaton.

A section in a quarry on Luddington Hill gives :—

|  | | | | ft. | ins. |
|---|---|---|---|---|---|
| Sandstone | - | - | - | 10 | 0 |
| Shale | - | - | - | 3 | 6 |
| Sandstone | - | - | - | 12 | 0 |
| Shale | - | - | - | 9 | 0 + |

Near Utkinton :—

|  | | | | ft. | ins. |
|---|---|---|---|---|---|
| Sandstone | - | - | - | 14 | 0 |
| Green and purple shale | - | - | - | 9 | 0 + |

In a quarry, near Tarporley, on the Chester Road :—

|  | | | | ft. | ins. |
|---|---|---|---|---|---|
| Flaggy sandstone | - | - | - | 29 | 0 |
| Shale | - | - | - | 6 | 0 |
| Sandstone | - | - | - | 10 | 0 |
| Shale | - | - | - | 5 | 6 |
| Flags | - | - | - | 3 | 0 + |

The flags and sandstones are worked up into window-sills.

The Waterstones contain ripple-marks abundantly, the casts and tracks of Annelids, sun-cracks, and so-called rain-pittings, the footprints of *Cheirotherium*, and the casts in clay of the cubical crystals of rock-salt. These casts or pseudomorphous crystals occur first in the Waterstones, being quite unknown in the Lower Keuper Sandstone. Cavities, containing crystals of carbonate of lime, are common ; they occur in both the red and the green beds, but the latter are more commonly found to be rich in lime. Footprints of *Cheirotherium* were first noticed by Sir Philip Egerton in flags supposed to have been obtained at Tarporley.* Others have been noticed at Kelsall.† They may still be obtained in many of the marl-pits on Luddington Hill. The sun-cracks and ripple-marks are preserved in these beds in great perfection, but the pittings which accompany them resemble in a few instances only those which are produced by rain. Indeed the frequency with which they occur in connexion with the sun-cracks points to so constant an alternation of sunshine and shower as to render this explanation of their origin suspicious. It is possible that in some cases they have been produced by the escape of gases from a moist surface of mud freshly exposed to a hot sun. Somewhat similar markings are produced by this cause in the tidal mud of the Mersey at every ebb tide.

The sharp division of the Waterstones from the sandstones and conglomerates beneath, and the strong lithological differences between them point to a complete change of conditions having taken place at this period. On the other hand, the Waterstones are closely allied to the Red Marls, and the passage from the one to the other is so gradual that it has not been found possible to draw a hard-and-fast line between them. The boundary which is dotted on the map is an approximate separation of the more sandy base from the main mass of overlying marls, the portion thus separated being about 200 feet in thickness.

---

* Proc. Geol. Soc., vol. iii., p. 14, 1835.
† Id., vol. iii., p. 100, 1839.

KEUPER; RED MARLS. FAULTS.      13

*The Red Marls* occupy a large area on the eastern side of the Quarter-sheet, but are much obscured by Glacial Drift. There are exposures at Newton, near Frodsham and Houndslow, where they are faulted against Waterstones. They are also exposed in a ravine running from Delamere Forest by Willow Wood to Mouldsworth, where they are thrown down by the great Delamere east and west fault. The beds consist of red and green shales with ripple-marks and numerous pseudomorphs of rock-salt. East of Eddisbury they are completely obscured by Drift Sand, but are probably faulted against the Basement Beds as indicated on the map. They are exposed again in marl-pits near Eaton, Tilston, and at the bottom of a sand-pit near the Canal-locks at Beeston, where they are nearly vertical. Near Peckforton they are faulted against Upper Mottled Sandstone.

They generally consist of shales and flags with a few thin sandstones. The rock-salt of Cheshire occurs in this subdivision, but is nowhere worked within the present district. Salt-pits in the township of Kingsley are noticed in the "Inquisitions of Edward the Third," but borings to a depth of 300 feet both here and at Acton have failed to discover rock-salt.*

## FAULTS.

The general run of the Faults in this area is from north to south, and their tendency is to counteract the effect of the easterly dip by throwing the beds down to the west, and repeating the outcrops in this direction.

A fault which appears to throw down the Pebble Beds of Waverton and Christleton is visible in the railway station at the former place. A small east and west fault was exposed in the cutting of the Cheshire Lines near Hoole, but from the prevalence of Drift the structure of most of the area occupied by the Bunter series must remain a matter of theory.

Exposures of faults are common in the Keuper area. The Helsby outlier is let down by a large fault, the position of which is fixed near Simmond's Hill by an exposure of Waterstones 100 yards west of the high road. The western boundary of this outlier is a fault with a downthrow to the east, a continuation of which is seen in the Manley Quarry, and in the woods south of the quarry. A small parallel fault traverses the east side of the quarry. A small fault with a downthrow of 12 feet to the west forms a cleft in the crags at Helsby, and a second traverses the centre of the outlier, but is nowhere visible. The sandstone which forms the roof of the "Caves" at the east end of the crags is thown down by a small fault, and a similar small outlier has been thrown down on the west side of Woodhouse Hill by a small fault which may be followed for a short distance in caves. The Keuper Marl boundary fault, west of Newton, is visible in a lane north of the boundary of this quarter-sheet. The break separating Simmond's Hill from Alvanley Cliff is due to a fault running east to Rough Hill, and a small branch of this fault may be seen near Lord's Well, running to the south-east. A north and south fault

---

* Ormerod, Quart. Journ. Geol. Soc., vol. iv., p. 262.

is visible at Rough Hill and another outside the Forest, 300 yards east of the preceding. It is cut off by an east and west fault visible at Kingswood, and running towards Wilkinson's Lodge on a line marked by the outbreak of springs. A fault throwing Keuper Basement Beds against Upper Mottled Sandstone is exposed in a quarry near Mouldsworth Station, and a north and south fault, before mentioned, is seen in the marl-pit at Irk Wood. A red marl boundary fault is exposed west of Waterloo.

On the south side of Delamere Forest the run of the strata and of the north and south faults is interrupted by the Delamere east and west cross-fault. It is exposed to view below the Forest Farm. Its effect is to throw down the Marls to the north, and, by introducing these soft beds between the harder Basement Beds of Simmond's Hill on the one side and Waterstones and Basement Beds of Longley and Eddisbury on the other, to give rise to a broad valley, which though partly filled up with Drift Sand, still forms a complete breach through the hills. A branch of the Delamere Fault is visible south-east of Eddisbury Lodge.

The westernmost of the four parallel faults of Kelsall is visible in a lane near that place. The positions of the others are easily ascertained from their throwing Waterstones against the Basement Beds. From the absence of any feature to mark the outcrop of the Basement Beds south of Kelsall, it is probable that they are cut off by one of these faults, as indicated on the map; but the rocks are concealed by Drift.

A second cross-fault runs from Cotebrook to Utkinton. It is nowhere visible, but its existence is proved by the relative levels of the Waterstones of Luddington Hill and the Basement Beds of Yew-Tree House. A third fault is exposed in a marl-pit east of Tarporley.

South of Tarporley the prevalent dip is to the south-east, and accordingly the Keuper Marls sweep round to the south-west, so as to underlie Beeston Brook. On the south side of the valley the crag of Beeston Castle with its capping of Basement Beds rises abruptly from the Drift-covered plain (see Frontispiece). There must, therefore, be a great fault between the marls of the valley and the sandstone-crag, throwing the marls down to the north not less than 600 feet. This fault is not actually visible, but probably runs close to a quarry situated north of the Dean Bank Road, where intensely shattered Keuper Sandstone, with infiltrated Heavy Spar (Sulphate of Barium), is exposed. The strong joints in the north cliff of the crag, which are lined with the same mineral, are parallel to this great fault.

The valleys of Beeston Brook and Delamere Forest are, therefore, similar in structure, for both owe their origin to the downthrow to the north of soft Keuper Marls by great east and west faults. From their great displacement and their effect on the physical features of the country, these faults may be considered to be the principal dislocations of the district.

The crag of Beeston Castle may be regarded as an isolated portion of the Peckforton escarpment, and it is interesting to find this small outlying fragment of Basement Beds exhibiting faithfully the distinctive physical features of this rock. The separation of this mass is attributable, indirectly, to the faults which enclose it on three sides. By their agency the patch of Basement Beds was thrown against softer beds to the south, east, and north. Long subsequently to the formation of these faults, denudation, acting along lines of weakness and removing the rocks at a more rapid rate in proportion as they were softer, was delayed at this fragment of tough Basement Beds, and left it standing, though diminished in size. It was therefore not by upheaval, but as a result of the degradation of the surrounding country that Beeston Crag came into existence.

The fault which throws the New Red Marl against Upper Mottled Sandstone is exposed near a lane leading up the hill from Peckforton.

From the prevalence of Drift in the area occupied by the Keuper Marls, and from the sameness of the beds, it has not been found possible to trace faults through them; but there is no reason to doubt that they are as much affected by dislocations as any of the other divisions of the Trias.

---

## PART II.

## SUPERFICIAL DEPOSITS.

### INTRODUCTION.

Out of a total area of 168 square miles in this Quarter-sheet more than 142 are covered by superficial deposits. As the character of these deposits is independent of that of the rock on which they rest, it will be seen how incomplete would be a survey in which their distribution was not observed. It has therefore been found necessary to publish a map* in which their various subdivisions are noted, as well as that showing the solid geology.

The following are found to occur :—

RECENT - - Alluvium. Fluviatile and Marine.

POST GLACIAL -⎰ Peat.
⎱ River Terraces.
⎱ Shirdley-Hill Sand.

GLACIAL - -⎰ Upper Boulder Clay.
⎰ Sands and Gravels.
⎱ Lower Boulder Clay (?).

### GLACIAL.

The Glacial Deposits form the continuation of those of Lancashire, where the three-fold division given above has been found to hold good over a large area. The Lower Boulder Clay of Lancashire is distinguished from the upper not only by position, but by certain marked characteristics, being generally of a more sandy texture, hard consistence, and more fully charged with erratic fragments.

*Lower Boulder Clay (?).*—It is not certain that any represen-

---

* Geological Survey Map, 80 S.W., showing the Superficial Deposits. The Drift deposits of the City of Chester and its suburbs were surveyed on the six-inch scale, and a copy of the map (38 Cheshire) has been deposited in the office of the Geological Survey, where it may be inspected or copied on application.

tative of this member of the Drift occurs within the present area, but in the following instances deposits resembling it in position and character have been observed.

Near the old training stables, south of Oakmere, there is a disused marl-pit, much overgrown, but in which the following section can be made out :—

|  | ft. |
|---|---|
| White sand, variable - | 2—4 |
| Laminated clay and red sand | 1—1½ |
| Red clay | 2—3 |
| Rocky red sand, with some clay, containing numerous pebbles and fragments of red sandstone, in part stratified - | 2+ |

There lay in this pit a boulder of Lower Keuper Sandstone measuring 3½ × 2 × 2 feet, of a grey colour, with a few pebbles. One side of the boulder was smoothed and marked by glacial striæ 6 or 8 inches long, and not glazed as in the case of slickenside. The quartzite pebbles had generally been splintered where they projected, but one had been finely polished and had protected the rock behind it, so as to give rise to a small ridge 1½ inches long.

I was informed[*] that during the construction of a sewer at Chester outside the Watergate and the City Walls south of it, a stony and rocky red clay was found to rest on the rock, and to be separated from an upper clay of the ordinary type by a line of sand or gravel.

In the railway-cutting at Newton near Chester, a thin seam of rocky red clay was observed[†] to underlie the drift sand and gravel and to rest directly upon the rock.

At Clough Brow, near Kingsley, there is seen in the bed of the stream a red clay underlying the sand and containing numerous fragments of Keuper Marl. It rests upon undisturbed Keuper Marls ; in the brook from the Slack to Kingsley there is red clay exposed under 45 feet of sand.

It must be remembered that the Boulder Clay frequently contains beds of sand interstratified with it. The occurrence of two clays, separated by a sand-bed, is therefore not sufficient to prove the existence of Lower Boulder Clay. But the deposits described above differ slightly from the ordinary Boulder Clay either in being largely made up of the débris of local rocks, or in containing a larger number of boulders.

It is a characteristic of this deposit that it occupies hollows in the rock-surface, so as to be very local in its distribution, and often overlapped in every direction by the Upper members of the Drift. Fresh sections may at a future date settle the question of its presence in this district.

The *Sands and Gravels* occupy an unusually large area in this sheet. They occur not only as a more or less continuous bed beneath the Upper Boulder Clay, but they rise through it to a considerable height in abrupt mounds and ridges. The ridges or sand-banks run in a definite direction, namely, from north-west to south-east, and almost always start from the south-east side of a rock-hill. They are most distinct on the western half of the

---

* By Mr. W. G. Shrubsole, F.G.S.
† By Mr. W. Shone, F.G.S.

district, where they sometimes extend across the Boulder Clay plain for as much as a mile and a half.

The great spread of sand which extends from Delamere eastwards beyond the margin of the sheet forms an undulating plateau, averaging 220 feet above the sea. The sand rises from under the Boulder Clay at Mouldsworth, runs through the valley of Delamere Forest, and spreads out in a broad fan-shape in the Keuper Marl area beyond, finally ending in an abrupt bank running from Kingsley by Stanney to Cuddington, and southwards by Budworth, Oulton, and Eaton. The surface of this great sand-bed is broken by mounds and ridges, the latter running parallel to the neighbouring rock features; the Long Ridge, for example, being parallel to the east face of Eddisbury Hill. Others appear to radiate eastwards from the Delamere Valley.

A similar deposit to that of Delamere, partly filled the Beeston Brook valley, and the remains of an old plateau, deeply cut into by the stream which flows through from the Boulder Clay plains on the east, but once continuous across the valley, are easily distinguishable on both sides. From Beeston and Tilston the sand runs to the east-south-east in a ridge nearly three miles long.

The form of the sand-banks and the distribution of the sand are such as would be produced by currents from the north-west. The check given to currents having this direction by such an obstacle as one of the Pebble Bed Hills would cause the deposition of sand under the lee of the hill, and give rise to a shoal or bank trailing away towards the south-east. The sand of Delamere and Beeston seems to have found its way under the influence of the same currents through the breaches in the hills at those places, and to have been deposited in the calmer water on the eastern side.

The surface of the sand of Delamere, Manley, Tilston, and Peckforton is variegated by numerous mere-basins and peat-filled hollows. Such features are characteristic of the Drift Sand and Gravel wherever it occupies a large area. The lakes either lie between the sand-ridges, or occupy hollows, which sink below the general level of the ground as rapidly as the sand-hummocks rise above it. They often have no streams running into them, nor any outlet; and the water in those which still exist as lakes is not held up by any impervious stratum, but stands at the water-level of the locality as regulated by the form of the ground and the depth of the neighbouring ravines.

It has been suggested that the mere-basins have resulted from the subsidence of the ground through the removal in solution of rock-salt in the Keuper Marls underlying the sand. The same features, however, are presented by the sand on whatever formation it occurs; the sand of Ellesmere, for example, lies on the Bunter, in which rock-salt is unknown. Moreover, in the Boulder Clay area, immediately east of the Delamere sand, there are no mere-basins, though it is nearer to the salt district and a more likely field for rock-salt. It is not probable that rock-salt occurs in the marls under Delamere Forest.

Q 4293.                                                    B

There can be no doubt that the main features of this tract, such as the sand-hummocks, the ridges with their intervening valleys, and the mere-basins, are surviving features of the original surface as it was left by the shifting currents by which the sand was distributed. The ravines which the existing streams occupy, though originally determined in direction by the sand-banks, have been excavated by subsequent erosion, and are proportional in depth and size to the amount of water flowing through them.

Sections of the Sand and Gravel are numerous in the pits where they have been dug for building purposes, and for road metal.

A bed of sand runs out from under Boulder Clay in the sides of a valley leading from near Chester to Backford and Stoke. A pit near Backford Hall shows :—

|  | | | | | ft. | ins. |
|---|---|---|---|---|---|---|
| Boulder Clay, thickening rapidly to the north | . | | . | . | 3 | 0 + |
| Sand, not continuous | - | - | . | - | . 2—3 | 0 |
| Tough chocolate clay, very regular | . | | - | . | 4 | 6 |
| Fine running sand, with veins of grey loam | | - | | . | . 20 | 0 + |

There are also several sections near Wervin, Caughall, and Moston, showing the superposition of the clay on the sand. One of the clearest is at Upton, where the accompanying sketch was made.

Fig. 3.

*Sand-pit at Upton.*

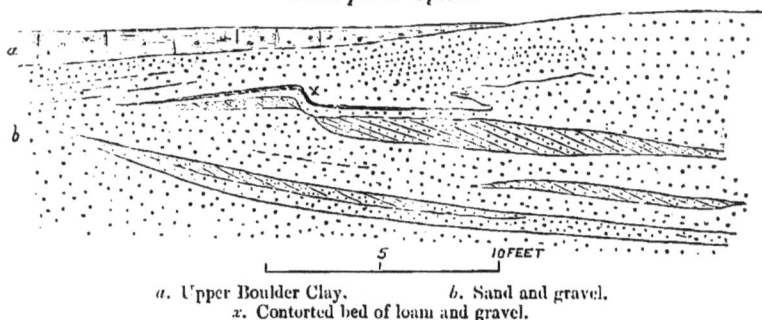

*a.* Upper Boulder Clay.          *b.* Sand and gravel.
*x.* Contorted bed of loam and gravel.

The Boulder Clay thins out to a feather-edge on an ascending slope of sand. The line of separation is sharp and clear, as is generally the case, but the bedding of the upper part of the sand seems to have been disturbed, and the displacement of the bed of loam and vein of shingle seem to have been the result of pressure from above.

At Upton Church the section is :—

| Boulder Clay - | - | - | 5 to 6 ft. |
|---|---|---|---|
| Sand - | . | - | . |

The sand rises to the surface again near the Cross Roads, at the windmill, and at Newton Hall, where it is of a more loamy texture than at Upton. It is exposed along the old course of the Flookersbrook through the nursery gardens by the General

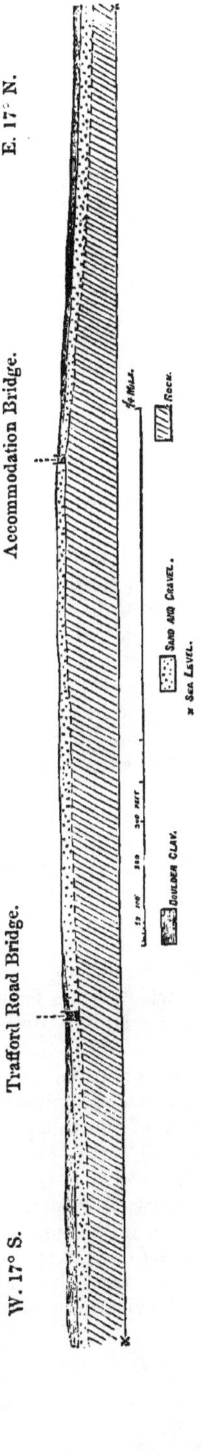

Railway Station to Hoole Park. The area lying between Upton, Newton, and the Flookersbrook may be generally described as being occupied by an undulating bed of sand, the shallow depressions in which are filled by Boulder Clay varying from 6 to 12 feet in thickness. The rock rises to the surface in the nursery gardens of Mr. James Dixon close to the office, but is not seen elsewhere within the above-named limits.

In a cutting near Hoole Bank in the Cheshire Lines, the accompanying section was exposed.(Fig. 4).

The upper part of the sand contains fine waving veins of loam, the lower part is sharp and loose. The same bed is exposed in the cutting of the London and North-Western Railway, and in the villages of Mickle Trafford and Guilden Sutton.

South of Hoole Park the sand rises from under the Boulder Clay and extends in a conspicuous bank to Christleton and thence to Waverton, connecting the rock-hills of those villages. The bottom of the sand was not reached at 48 feet at Rowton; at Christleton it was 8 feet thick and rested on the rock. It is fine grained, with a few veins of loam, but no gravel. The superposition of the Boulder Clay is shown in various places near the waterworks along the Tarvin Road and in the Chester and Crewe Railway. Springs are thrown out along the junction.

Similar ridges extend from the south-east sides of Barrow, Tarvin, and Stapleford hills, and after being overspread by Boulder Clay for a short distance, re-appear at Duddon Common and Clotton. The same bed is exposed in the ravines leading into the Gowy near Stapleford Mill. The accompanying sketch was made on the west side of the river near the high road.

FIG. 4.

Section in the Cheshire Lines, near Hoole.

W. 17° S.    Trafford Road Bridge.    Accommodation Bridge.    E. 17° N.

ROCK.    SAND AND GRAVEL.    SEA LEVEL.    BOULDER CLAY.

A bed of Boulder Clay, 18 inches thick, occurs in the sand at Sandy Lane, near Burton. It dips towards the north and thickens in the same direction.

FIG. 5.

*Sand-pit at Stapleford.*

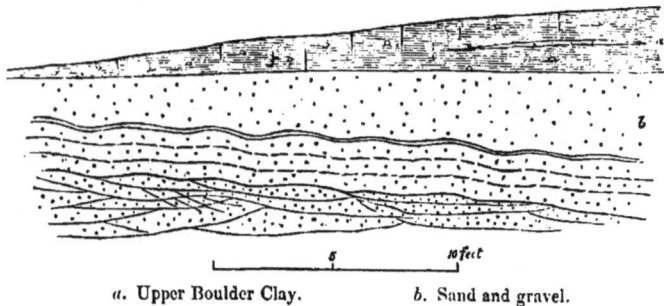

*a.* Upper Boulder Clay.              *b.* Sand and gravel.

West of Tarporley a barrier of sand extends across the mouth of a shallow valley in the Waterstones, and has given rise to a mere-basin now filled with peat. The sand is interbedded with clay in thin veins and bands of one or two feet in thickness. The whole is contorted, and the sand and clay jumbled up together.

East of Manley and Alvanley there is a considerable area occupied by drift sand, bounded on the east by hills of the lower Keuper Sandstone. The Ridgeway is a bank of sand running from N.W. to S.E., with seams of gravel and fragments of shells of *Turritella* and *Cardium*. Outlying patches of Boulder Clay, remnants of a once more extensive sheet, occur near Alvanley and Higher Hall. A pit near the high road one mile south-east of Alvanley, shows beds of gravel made up of 35 per cent. of local Red Sandstone, some rolled lumps of Boulder Clay, a few granite boulders, and a large number of Lake District erratics; the gravel beds are interstratified with and wedge into beds of fine sand, containing small fragments of shells. There are four small peat-mosses and two alluvial flats near Higher Hall, all occupying the sites of former mere-basins in the sand. The underlying rock is Waterstones. The large quantity of water which issues from the sides of the valley on the north side of Higher Hall, is probably held up in the sand by these beds.

Towards the south, this deposit of sand is connected with the great spread of Delamere Forest. The most westerly exposure is near Peel Hall, where it runs along the sides of the valley under about 3 to 6 feet of Boulder Clay; there is the usual appearance of disturbance in the upper part of the sand.

From this point the level of the ground rises to Mouldsworth and Ashton Heys, the Boulder Clay extending up portions of the slope as far as the edge of the Forest. It must once have formed a continuous sheet over the sand at least as far as this, but its margin has been cut back into its present labyrinthine form by the numerous springs which break out of the sand, and have excavated deep ravines leading down to the main valley.

The distribution of the beds of pebble gravel in the sand is very irregular, but as a rule they are more frequently found, and the sand is coarser in grain, in the neighbourhood of the hills than in the plains.

A gravel-pit by the side of the Norley and Mouldsworth high road near the west end of Blake Mere, shows beds of fine grit and sand interstratified with shingle beds containing a few fragments of shells. The larger pebbles average 3 × 2 × 1 inches; they are all well rounded. On to the east side of the high road from Delamere Station to the Abbey Arms, there is a gravel-pit containing about 60 per cent. of local red sandstone, the remainder Lake District erratics with a few granites.

At Peel Hall, west of Kingsley, there are beds of gravel in the upper part of the sand, and west of Norley there are a few irregular bands near the surface worked for road metal. A gravel-pit near Delamere House shows a little fine sand overlying gravel. The pebbles in the latter average $1\frac{1}{2}$ × 1 × $\frac{1}{2}$ inches, and 10 per cent. only are local, a smaller percentage than in the gravel nearer the hills. The current-bedding is to the E.S.E. A brook section at the Slack near Kingsley shows the sand to be 45 feet thick, resting on red clay. The sand at Stanney Brook contains very little gravel, but occasional subangular masses of Boulder Clay. These masses are probably the fragments of a bed once continuous and interbedded in the sand, but afterwards displaced, for the bedding of the sand in which they occur bears evidence of having been forcibly disturbed. The accompanying sketch of a similar displacement was taken in a sand-pit at Beech Moss, near Norley.

Fig. 6.

*Sand-pit near Beech Moss.*

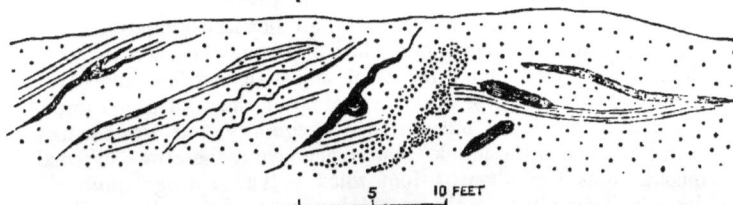

Showing contorted bedding.

It is certain that these beds of fine loam, clay, and fine shingle cannot have been deposited by water in the positions which they now occupy. At Tarporley, again, beds of sand interstratified with clay in beds of one to two feet in thickness are in the same unnatural position. A pit at Beeston shows a bed of clay with boulders, once evidently continuous, but now in squeezed and disunited fragments; and at Guilden Sutton (Fig. 7, p. 26) may be seen pockets of sand in boulder clay, probably the remains of a crumpled-up sand-bed. The displacement of the bed of loam at Upton (Fig. 3, p. 18) and the obliteration of the bedding of the upper

part of the sand in the sections at Stapleford and Peel Hall, before
mentioned, are similar instances of disturbance. Such disturbances
seem to point to the action of ice. In the case of the intensely
contorted drift of the south-eastern counties, it has been suggested
that they may have been caused by the imbedding of masses of
ice or frozen mud which afterwards melted. In other cases they
have been accounted for by the stranding of heavy masses of floating
ice on sand-banks. It is noticeable that they are found only in
connexion with those deposits which were formed by the direct
agency of ice, namely, the Boulder Clay. The beds of sand and
gravel, when not associated with Boulder Clay, are undisturbed,
and except in the indirect evidence of erratics, which were probably
derived from the waste of pre-existing Boulder Clays, show no trace
of the action of ice.

Two beds of Boulder Clay occur in the drift sand at Cuddington
Waste, where a descending section shows:—

| | feet. |
|---|---|
| Strong chocolate-coloured clay, with few boulders    - | 10 |
| Sand, the upper part with contorted veins of chocolate-coloured clay, the lower part fine and current-bedded and in places *faulted* as if by pressure   -    -    - | 8—10 |
| Strong chocolate-coloured clay, best marl    -    - | 8 + |

This pit is open to all inhabitants of Cuddington to obtain marl
for land. From the scarcity of clay in the district, the beds are
especially valuable. At Cuddington Station, the bottom of the
sand was not found at 60 feet in an unsuccessful search for
water.

Another valuable bed of clay was met with in constructing the
reservoir of the Winsford Waterworks, near Shaw Brook. This
brook has its rise in a spring near the Long Stone, and is fed by
numerous others issuing from the sides of the valley which it has excavated in the sand. It is joined near Clay Lane Farm by a smaller
brook fed in the same way. It having been determined to impound
the springs feeding Shaw Brook for the supply of Winsford, a reservoir was constructed in the bottom of the valley nearly opposite Shaw
Farm, and in the excavations a bed of strong chocolate-coloured
Boulder Clay more than 9 feet thick and with very few stones was
found. It lay under or near the bottom of the sand which forms
the sides of the valley, but ran very irregularly, and was not met
with lower down the brook. Two feet above it there was a second
incontinuous bed about 1 foot thick. The springs probably
draw their supply from the area lying round Sandy Brow, Folly
Farm, and part of Newchurch Common. From the porosity of the
soil nearly the whole of the rainfall is absorbed; the excess over
what the sand can retain breaks out in the ravines in the form of
springs, the position of which may be determined by the accidental
presence of beds of clay. In a well at Sandy Way, near Oakmere,
water was met with at 75 feet, but the bottom of the sand not
found at 78 feet; and at Mr. Leather's Farm, near the Chester
and Northwich Road, a quicksand was found at 36 feet; a thin bed
of marl was once worked on the north side of Relick's Moss. At
Outside Farm the sand contains a seam of clay 2 inches thick,
dipping to the N.W. at 10°.

South of Budworth the breadth of the sand area rapidly narrows, so that near Eaton the Boulder Clay overlaps it and rests on the marls. But at Hickhurst Lane the sand rises abruptly in a dome-shape through the Boulder Clay. In a sand-pit it is seen to contain a bed of clay about 1 foot thick, and to be inter-stratified with a little gravel, the larger pebbles averaging $1\frac{1}{2} \times \frac{3}{4} \times \frac{1}{2}$ inches; the whole of the beds slope to the west, and the clay thickens in the same direction. The dip of the current-bedding is towards the E.S.E. at the Rabbit Burrows, near Tilston Hall, but towards the N.W. at Mill Hall and Outside Farm.

From Tilston Fernall the sand again stretches out into the Boulder Clay plain to Calveley Park, where it is seen in a sand-pit to be fine and reddish, without gravel. At the former place it is fine with the usual wavy veins of loam, and with a thin bed of marl dipping to the W.N.W.

In the valley of Beeston Brook the Keuper Marls are exposed under the Drift Sand, in a pit near the canal locks. The sand is fine-grained to the base, and is only separated from the Keuper Marls by a few inches of red clay. Both here and at Tiverton it is interbedded with loamy bands in a nearly horizontal position. It differs from the sand which occupies a similar position in the Delamere Valley in being almost free from gravel.

The area between Beeston, Bunbury, and Peckforton is over-spread by sand of no great thickness, through which Keuper Marls are occasionally exposed. Out of several drift-hollows in this area, that of Peckforton Mere alone contains water. There is more gravel here than in the sand of Tilston and Calveley. Coarse gravel containing about 60 per cent. of local rocks, chiefly Lower Keuper Sandstone, is interstratified with coarse sand at Beeston Gate, the current-bedding planes being inclined to the south. A sand-pit at Peckforton shows fine sand with loamy bands, and is occasionally current-bedded with gravel containing 50 per cent. of local rocks, chiefly Lower Keuper Sandstone, with the small quartzite pebbles which occur in it.

The sand extends eastwards to Hall Green, where it is inter-stratified with shingle, containing only 10 per cent. of local fragments. The current-bedding planes are towards the east-south-east.

Fragments of shells are generally to be found in the Sands and Gravels in all the above localties.* They are commonest in the finer bands in the gravel beds, but they are invariably fragmentary and waterworn. A list of those found by Mr. Shone at Upton is given hereafter; the usual species are *Cardium edule*, *Turri-tella terebra*, and *Tellina balthica*.

The *Upper Boulder Clay* occupies the largest area of all the formations in this Sheet. Its limits are determined by the form of the ground irrespectively of the height above the sea. They,

---

* Shells were first noticed in the Drift Gravels of this district by Mr. J. Trimmer in 1883; Proc. Geol. Soc., vol. i., p. 419. And subsequently at Willington and Norley Bank by Sir P. de M. G. Egerton; Proc. Geol. Soc., vol. ii., pp. 189 and 415, 1835-6.

moreover, do not run along any given level, but run up to a
greater elevation on the side of a hill where the slope is gentle
than where it is steep. As a general rule this deposit is thickest
in the lowest ground, so that by filling up hollows it tends to
obliterate or modify the inequalities of the surface of rock or
sand on which it rests.

It consists of a red or chocolate-coloured clay, with occasional
traces of stratification indicated by incontinuous veins of sand,
lamination, or the arrangement of boulders along definite lines.
But more commonly the boulders are distributed irregularly
through the deposit and imbedded in every possible position. It
is seen in the brick-pits to break more readily along vertical
joints than horizontal planes. These joints extend to a depth of
6 or 8 feet from the surface, and are lined by a thin green film of
protoxide of iron, the result of the reduction of the red peroxide
by soil-water, charged with the products of decomposition of
vegetable matter. Erratic blocks of volcanic, granitic, and
metamorphic rocks are found everywhere in this deposit, and
commonly show the striations characteristic of the action of ice.
They decrease, however, both in size and abundance in proceeding
from north to south ; that is, in receding from the source of supply.

The superposition of the Boulder Clay on the Sand and Gravel
has been repeatedly noticed in the preceding pages. Though the
clay is often seen to rest at its margin directly on the rock, it is
usually found in sinking wells to be underlain by sand, more
especially in low ground, indicating that the sand is overlapped
where the Drift abuts against a rising surface of rock. This is
the case under Chester ; the city stands upon the rock, but towards the lower ground a thin covering of Boulder Clay appears,
commencing at the Eastgate, and thickening eastwards. Near
the City Road a thin wedge of sand comes in, separating the
clay from the rock, and developing rapidly into the great deposit
which rises to the surface at Boughton. Beds of sand in such a
position are usually full of water, which is held down in them
under pressure by the impervious covering of clay.

The general observations given above are illustrated by the
accompanying sections, which have been supplied to me by the
kindness of all those from whom I have had occasion to seek for
information :—

THE CEMETERY, CHESTER :—      feet.
    Boulder Clay  .  .  .  .  . 20
    Rock  .  .  .  .  . —
BANKS OF THE DEE, UNDER CURZON PARK :—
    Boulder Clay  .  .  .  .  . 30
    Sand  .  .  .  .  . 5
    Rock  .  .  .  .  . —
SELLER'S BREWERY, CHESTER :—
    Boulder Clay  .  .  .  .  . 18
    Rock  .  .  .  .  . 30 (+)
LEAD WORKS, CHESTER :—
    Boulder Clay  .  .  .  .  . 12
    Sand  .  .  .  .  . 1½
    Gravelly clay  .  .  .  .  . 3
    Sand, full of water  .  .  .  .  . —

feet.

QUEEN'S HOTEL, CHESTER :—
Boulder Clay . . . . . 7
Gravel, with a little water - . . . —
Clay . . . . . . —
FLOOKERSBROOK, CHESTER :—
Boulder Clay . . . . - 12
Sand . . . . . —
BISHOP'S FIELDS, CHESTER :—
Boulder Clay . . . . - 1
Loose sand . . . . - 60 (+)
TARVIN ROAD, NEAR THE CANAL BRIDGE :—
Boulder Clay . . . . - 6
Sand . . . . . —
THE RAKE, ECCLESTON :—
Boulder Clay . . . . - 24
Rock . . . . . —
UNDER THE NEW TOWER, EATON HALL :—
Boulder Clay . . . . - 90 (+)
EATON HALL WATERWORKS BY THE RIVER :—
Boulder Clay . . . . - 20
Rock . . . . . - 327 (+)
BROOK HALL, TATTENHALL :—*
Boulder Clay, about . . . - 60
Sand . . . . . —
BANK HOUSE, TATTENHALL :—†
Boulder Clay . . . . - 50
COTTON ABBOTS, TATTENHALL :—*†
Boulder Clay . . . . - 48
Sand . . . . . —
HALF MILE S. OF NEWBOLD LODGE, BRUERA :—*
Boulder Clay . . . . - 36
Sand . . . . . —
HALF MILE S. OF HATTON HEATH :—
Boulder Clay . . . . - 50 (+)
MOLLINGTON :—*
Boulder Clay . . . . - 7
Sand . . . . . - 5
Boulder Clay . . . . - 47
Sand with water . . . . —
PICTON HALL, MICKLE TRAFFORD :—*
Boulder Clay . . . . - 15
Sand with water . . . . —
WHITBY :—
Boulder Clay . . . . - 20
Sand . . . . . - 70
Rock . . . . . —

In several of the above sections beds of sand of no great thickness occur in the Boulder Clay; they are not continuous and do not afford much water. The clay immediately over or under such beds is laminated, and assumes the character of a reddish brown loam. The beds of laminated clay passed through in two of the wells at Tattenhall were no doubt of this character, and might be found to develop into beds of sand. A thin seam of sand can be seen in the clay at Stapleford (Fig. 5, p. 20) in an undisturbed position, but in other cases the sand occurs in pockets in the Boulder Clay, as in the Guilden Sutton sand-pit.

* The water rose high in the well as soon as the clay was penetrated.
† Beds of laminated clay were passed through. A little water is sometimes found under such beds.

Such pockets are probably the remains of a sand-bed originally interstratified but subsequently jumbled up.

Fig. 7.

*Sand-pit at Guilden Sutton.*

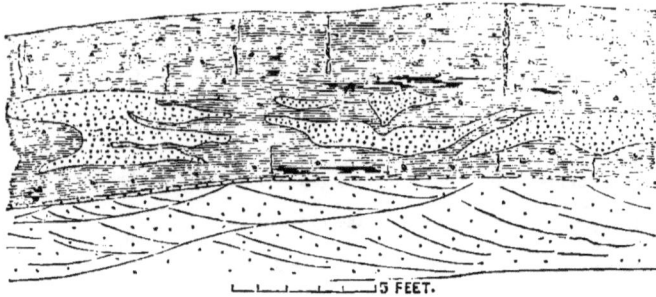

5 FEET.

*Fossils.*—The Boulder Clay is generally found to contain marine shells. They occur as broken and waterworn fragments, and are obviously as little in place as the boulders themselves.* Mr. W. Shone, F.G.S., in collecting from the clay of the Newton cutting near Hoole, observed that the cavities of the *Turritellæ* were often filled with a blue silt, different in appearance to the red clay in which the shells were imbedded. An examination with the microscope revealed the presence in the blue silt of numerous *Foraminifera* and *Ostracoda.*

The following lists of the *Foraminifera* and *Mollusca* in the sand and gravel of Upton and the Upper Boulder Clay of the Newton Railway cutting are extracted from Mr. Shone's papers:—†

FORAMINIFERA.

Cornuspira involvens, *Philippi.*
Biloculina ringens, *Lamk.*
—— elongata, *D'Orb.*
Triloculina trigonula, *Lamk.*
—— oblonga, *Montagu.*
Quinqueloculina seminulum, *Linn.*
—— bicornis, *W. & J.*
—— secans, *D'Orb.*
—— subrotunda, *Montagu.*
—— agglutinans, *D'Orb.*
—— Ferussacii, *D'Orb.*
Lituola scorpiurus, *Mont.*
—— canariensis, *D'Orb.*
Lagena sulcata, *W. & J.*
—— lævis, *Montagu.*
—— striata, *D'Orb.*
—— semistriata, *Will.*
—— globosa, *Montagu.*
—— marginata, *W. & J.*
—— lucida, *Will.*

Lagena squamosa, *Montagu.*
Nodosaria scalaris, *Batsch.*
—— radicula, *Linn.*
—— pyrula (?), *D'Orb.*
Dentalina communis, *D'Orb.*
Cristellaria rotulata, *Lamk.*
—— crepidula, *F. & M.*
Polymorphina communis, *D'Orb.*
—— lactea. *W. & J.*
—— compressa, *D'Orb.*
Uvigerina angulosa, *Will.*
Orbulina universa, *D'Orb.*
Globigerina bulloides, *D'Orb.*
Textularia variabilis, *Will.*
—— globulosa, *Ehrenb.*
—— pygmæa, *D'Orb.*
—— difformis, *Will.*
Bulimina pupoides, *D'Orb.*
—— marginata, *D'Orb.*
—— aculeata, *D'Orb.*

* William Smith, the "father of English Geology," is said to have noticed the occurrence of shells in the clay-pit at Saltney.

† Quart. Journ. Geol. Soc., vols. xxx. and xxxiv., and Proc. of the Chester Soc. Nat. Sci., No. 2.

## FORAMINIFERA—continued.

Bulimina ovata, *D'Orb.*
—— elegantissima, *D'Orb.*
—— spinulosa (?), *Will.*
Virgulina Schreibersii. *Crjzck.*
Bolivina plicata, *D'Orb.*
Cassidulina lævigata, *D'Orb.*
—— crassa. *D'Orb.*
Discorbina rosacea, *D'Orb.*
—— globularis, *D'Orb.*
Planorbulina mediterranensis, *D'Orb.*
Truncatulina lobatula, *Walker.*

Truncatulina refulgens, *Mont.*
Pulvinulina repanda, *F. & M.*
Rotalia Beccarii, *Linn.*
—— nitida, *Will.*
Patellina corrugata, *Will.*
Polystomella crispa, *Linn.*
—— striato-punctata, *F. & M.*
Nonionina umbilicatula, *Montagu.*
—— depressula, *W. & J.*
—— asterizans, *F. & M.*

## OSTRACODA.

Cythere pellucida, *Baird.*
—— tenera, *Brady.*
—— finmarchica, *G. O. Sars.*
—— villosa, *G. O. Sars.*
—— concinna, *Jones.*
—— tuberculata, *G. O. Sars.*
—— Dunelmensis, *Norman.*
—— Whitcii, *Baird.*
—— antiquata. *Baird.*
—— Jonesii, *Baird.*
Cytheridea papillosa, *Bosquet.*
—— punctillata, *Brady.*
—— Sorbyana, *Jones.*

Eucythere argus, *G. O. Sars.*
Loxoconcha impressa, *Baird.*
—— guttata, *Norman.*
—— tamarindus, *Jones.*
Cytherura striata, *G. O. Sars.*
—— angulata, *Brady.*
—— producta, *Brady.*
Cytheropteron latissimum, *Norman.*
—— nodosum, *Brady.*
Sclerochilus contortus, *Norman.*
Paradoxostoma ensiforme, *Brady.*
—— flexuosum, *Brady.*
—— arcuatum, *Brady.*

In the tables of Mollusca v r means that one to three specimens have occurred; r, three to ten; f, frequent; c, common; a, abundant; v, very.

LIST of MOLLUSCA, &c., from the DRIFTS of WEST CHESHIRE.

| | Middle Sands and Gravels, Upton. | Upper Boulder Clay, Newton-by-Chester. |
|---|---|---|
| Anomia ephippium, *Linn.* - - - - | — | v r |
| Ostrea edulis, *L.* - - - - | r | f |
| Pecten opercularis, *L.* - - - | r | f |
| Mytilus edulis, *L.* - - - - | | f |
| —— modiolus, *L.* - - - - | r | f |
| Nucula nucleus, *L.* - - - - | — | v r |
| Leda pernula, *Müller* - - - | — | v r |
| Pectunculus glycymeris, *L.* - - - | — | f |
| Area lactea, *L.* - - - - | — | v r |
| Cardium echinatum, *L.* - - - | c | a |
| —— edule, *L.* - - - - | a | a |
| Cyprina islandica, *L.* - - - | f | a |
| Astarte sulcata, *Da Costa* - - - | r | r |
| —— sulcata, var. elliptica - - - | — | f |
| —— compressa, var. striata - - - | — | v r |
| —— borealis, *Chemnitz* - - - | r | a |
| Venus lincta, *Pulteney* - - - | — | v r |
| —— chione, *L.* - - - - | — | r |
| —— casina, *L.* - - - - | — | v r |
| —— gallina, *L.* - - - - | — | f |
| Tapes virgineus, *L.* - - - | — | v r |
| Tellina balthica, *L.* (T. solidula, *Pult.*) | a | v a |
| —— calcaria (T. proxima, *Brown*) - - | — | f |
| Psammobia ferröensis, *Ch.* - - - | r | f |
| Mactra solida, *L.* - - - - | r | — |
| —— solida, var. elliptica - - - | — | c |
| —— subtruncata, *Da C.* - - - | — | v r |

LIST OF MOLLUSCA, &c.—*continued.*

| | Middle Sands and Gravels, Upton. | Upper Boulder Clay, Newton-by-Chester. |
|---|---|---|
| Thracia pubescens, *Pult.* - - - - | -- | v r |
| Corbula gibba, *Olivi* (C. nucleus, *Lam.*) - | -- | v r |
| Mya truncata, *L.* - - - - - | r | c |
| Saxicava rugosa, *L.* - - - - - | — | r |
| —— rugosa, var. arctica - - - - | — | r |
| —— crispata, *L.* - - - - - | — | v r |
| Dentalium entalis, *L.* - - - - | — | f |
| —— striolatum, *Stimpson* (D. abyssorum, *Sars*) - | v r | f |
| Trochus cinerarius, *L.* - - - - | r | — |
| Lacuna divaricata, *Fabricius* - - - - | — | r |
| Littorina obtusata, *L.* - - - - | -- | v r |
| —— rudis, *Maton* - - - - - | — | f |
| —— litorea, *L.* - - - - - | — | c |
| Homalogyra atomus, *Philippi* - - - - | — | v r |
| Turritella terebra, *L.* - - - - | a | v a |
| Scalaria communis, *Lam.* - - - - | — | v r |
| Odostomia interstincta, *Mont.* - - - | -- | v r |
| Natica sordida, *Ph.* - - - - | — | v r |
| —— catena, *Da C.* - - - - | v r | — |
| —— Alderi, *Forbes* - - - - - | — | v r |
| —— affinis, *Gmelin* (N. clausa, *Broderip & Ponsonby*) | -- | v r |
| Admete viridula, *Fabricius* - - - - | — | v r |
| Aporrhais pes-pelecani, *L.* - - - - | r | v r |
| Purpura lapillus, *L.* - - - - - | f | f |
| Buccinum undatum, *L.* - - - - | f | f |
| Murex erinaceus, *L.* - - - - | f | f |
| Trophon clathratus, *L.*, var. truncata - - - | r | c |
| Fusus antiquus, *L.* - - - - - | — | v r |
| Nassa reticulata, *L.* - - - - | f | r |
| Pleurotoma rufa, *Mont.* - - - - | — | v r |
| —— turricula, *Mont.* - - - - | — | f |
| —— pyramidalis, *St.* - - - - | — | f |
| Cypraea europaea, *Mont.* - - - - | — | v r |

POLYZOA.

| | | |
|---|---|---|
| Salicornaria Cuvieri, *Lam.* - - | — | r |
| Lepralia Peachii, *Johnston* - - - | — | v r |

CIRRIPEDIA.

| | | |
|---|---|---|
| Balanus crenatus, *Brug.* - - - - | -- | v r |
| —— sulcatus, *Lam.* - - - - - | — | v r |

ANNELIDA.

| | | |
|---|---|---|
| Serpula vermicularis, *Ellis* - - - - | — | v r |
| Spirorbis nautiloides, *Lam.* - - - - | — | r |

ECHINOIDEA.

| | | |
|---|---|---|
| Cidaridae ?, spines of, from the sand within Gastropoda | -- | f |
| Spatangidae, spines of, from the sand within Gastropoda | — | f |
| Toxopneustes drobachiensis, *Müll.* - - - | — | v r |

SPONGIDA.

| | | |
|---|---|---|
| Cliona, sp., in shells - - - - - | r | r |
| Grantia, sp., spicula of, from the sand within Gastropoda - - - - - - | | f |

*Boulders.*—In addition to the Boulders which are imbedded in the Drift, several occur on the surface of the ground at all elevations. Though numbers have been broken up, they still are fairly abundant. The following are the principal :—

| | | ft. | ft. | ft. |
|---|---|---|---|---|
| Dunham Hill - - - | Grey Granite - - - | 2 | × 2 | × 3 |
| Great Barrow - - - | Lake District Volc. Series - | 5 | × 4 | × 2 |
| „ - - - | „ „ - | 3 | × 3 | × 2 |
| Golbourne Bridge, Handley - | Grey porphyritic Granite - | 2¼ | × 1½ | × 1 |
| Shepherd's House - - | Grey Granite - - - | 4 | × 2 | × 2 |
| „ - - | „ - - - | 3½ | × 3¼ | × 2 |
| Riley Bank, near Alvanley - | Lake District Volc. Series - | 4 | × 2 | × 2 |
| „ „ - | „ „ - | 3½ | × 3 | × 2 |
| Newton, near Frodsham - | Grey Granite „ - - | 5¼ | × 4 | × 2 |
| Maiden's Cross - - | „ - - - | 3 | × 2½ | × 2 |
| Ashton - - - | „ - - - | 2½ | × 1 | × 2 |
| Duddon Heath - - | Volcanic Ash - - - | 2½ | × 2 | × 1 |
| Duddon - - - | Lake District Volc. Series - | 3 | × 2½ | × 1 |
| Willington Mill - - | Grey Granite - - - | 4 | × 3 | × 2 |
| Rook House - - - | „ - - | 3 | × 2 | × 2 |
| „ - - - | Lake District Volc. Series - | 5 | × 2 | × 1 |
| Heald, Eddisbury - - | Grey Granite - - - | 3½ | × 2 | × 2 |
| Delamere Church - - | Fine Grey Granite - - | 3½ | × 3 | × 3 |
| „ - - | Coarse Grey Granite - - | 6 | × 3 | × 2 |
| „ - - | „ „ - - | 4 | × 3¼ | × 2 |
| „ - - | Porphyry - - - | 3 | × 2½ | × 1½ |
| Eddisbury Hill - - | Lake District Volc. Series - | 10 | × 5 | × 4 |
| Norley - - - | Grey Granite - - - | 3½ | × 2 | × 2 |
| Cuddington Waste - - | Lake District Volc. Series - | 4 | × 4 | × 4 |
| Oulton Hall, Budworth - | Grey Granite - - - | 4 | × 2 | × 1 |
| Eaton, half mile south of - | Lake District Volc. Series - | 2 | × 2 | × 1 |
| „ „ - | Grey Granite - - - | 4 | × 3 | × 1 |
| Tilston Hall - - - | „ - - - | 4 | × 3 | × 2 |
| „ - - - | „ - - - | 6 | × 5 | × 4 |
| „ - - - | Five blocks of Grey Granite, averaging - - - | 2 | × 2 | × 1 |
| „ - - - | Pink Granite - - - | 2 | × 2 | × 1 |
| „ - - - | Ten blocks of Lake District Volc. Series, averaging - | 2 | × 1 | × 1 |
| „ - - - | Porphyry - - - | 1½ | × 1½ | × 1 |
| Rowton - - - | Keuper Sandstone (glaciated) - | 2¼ | × 1½ | × 1½ |
| Training Stables, near Oak Mere | „ „ „ - | 2½ | × 2 | × 2 |

It will be noticed that a large number of the Boulders have come from the Lake District. The granites are probably Scotch.

*Terminal Curvature.*—In the valley of the Mersey and the districts to the north of the present area, the rock frequently shows the characteristic smoothing and striation due to ice, especially where its surface has been protected from the weather by a covering of Boulder Clay. Though no instances of glaciated surfaces have been met with in this area, partly owing to the rock being of an unsuitable nature to retain the markings and partly from absence of sections, there is evidence of the passage of a heavy body over the ground, in the bending and drawing out of soft beds as if by pressure. The accompanying sketch was taken in a pit in the Keuper Marls at the north-east corner of Castle Hill Wood. The beds have not only been bent over at at the surface, but have been dragged in a north-easterly direction

for a distance of 8 feet or more. There is no slope in this
direction to account for the movement by the slipping of the
surface.

FIG. 8.

*Marl-pit at the N.E. corner of Castle Hill Wood.*

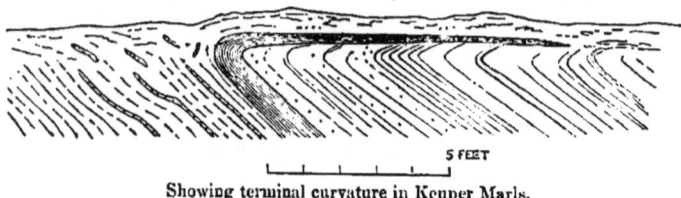

5 FEET

Showing terminal curvature in Keuper Marls.

## POST-GLACIAL AND RECENT.

*Shirdley Hill Sand.*—This is a deposit of subaërial formation,
extending from the valley of the Mersey northwards. It was
first described by Mr. De Rance in 1870,* and named from a hill
near Ormskirk which was considered by him to mark a line of
ancient sand dunes, formed at a time when the land stood at a
lower level than at present. The most southerly outliers of this
deposit occur within this area, one north of Wimbold Trafford, the
other near Hapsford. Both occur at the margin of the tidal
alluvium, and are partially overgrown by peat. The sand occurs
as a thin irregular bed lying on an eroded surface of Upper
Boulder Clay. It is quite free from gravel. The best sections
are to be found in the cliffs on the north bank of the Mersey
between Hale and Garston.†

*River Terraces.*—The small portion of the Weaver Valley
which comes within this Sheet shows well-preserved terraces of
gravel at a level of 15 to 20 feet above the present level of the
river.

The gravel is coarse and interstratified with sand or grit; on
the one side it rests against a bank of Keuper Marl, which forms
the original side of the valley, while on the other it ends off in
a bank, from the foot of which extends alluvium of recent
formation.†

*Peat.*—The only peat beds of any importance occur in the
mere-basins in the Drift Sand. Many of these, which were
filled with water within the memory of old inhabitants, are now
dry, partly owing to the deposition of silt or the accumulation of
vegetable matter,‡ partly by the construction of artificial outlets
for the water. Out of a total of 62 in the Delamere district, four
only still contain water. Of these the largest is Oakmere,
measuring 1,200 yards in length by 330 in its greatest
breadth. There are no streams running either into or out of it,

* "Geology of Liverpool and Southport," *Geol. Surv. Mem.,* p. 5. See also same
author, "Superficial Deposits of South-west Lancashire," *Geol. Surv. Mem.,* 1877,
p. 58.
† See Explanation, Quarter-sheet 80 N.W., Ed. 3, in press.
‡ I was informed by Mr. Leather, of Delamere, that large trunks of oak and
Scotch fir were met with in draining the mosses on his farm.

but in wet weather water escapes by soaking through the sand, and breaks out as a spring 300 yards beyond the north end of the mere.

The largest of the peat mosses is Blake Mere, 1,100 yards by 650 at the broader end. The black soil and dark foliage of the firs in these swampy hollows form a striking contrast with the sunny sand-banks surrounding them.

*Fluviatile and Tidal Alluvium.*—Hapsford Moor and the lower part of the valley of the Gowy are occupied by laminated silt, left by the ebbing tide, and locally known as " Slutch." Since the erection of cops and sluices, the marshes are no longer overflowed by the tide, and the deposition of slutch has ceased. It is the most modern of the river deposits, and inland shades insensibly into the fluviatile alluvium. At Ince and Frodsham it overlies a prehistoric forest bed (Explanation 80 N.W.).

The windings of a river through alluvial flats are subject to great changes. In a sketch plan of Chester of the date 1574, the river Dee is represented as turning more rapidly than at present after passing the Castle, so as to flow past the Watergate and along the Walls to the Water Tower, and as spreading out thence into a broad channel along the course now known as Finchett's Gutter. In consequence of the continued silting up of this broad channel, the existing artificial course from Chester to Saltney and thence to Connah's Quay was made by the Dee Navigation Company in the years 1733–6, and this course has since been maintained by protecting the banks with cheverils. But the change in the course from the Water Tower to the Castle appears to have been due to natural causes, and to have resulted from the undermining of the banks on the outer side of a curve, and the simultaneous silting up the inner side to the flood-level, a process which takes place in all rivers. A large part of the Rood Eea has therefore come into existence since the date of this plan of the city. The alluvium of the Rood Eea and Sealands consists of fine sand and loam with marine shells.*

Where the river valley is excavated in the rock, it is narrow, as at Chester and Heron Bridge; but in the Boulder Clay it widens out, from the ease with which the banks have yielded to the winding of the stream. Between Boughton and Heron Bridge the alluvium is nearly half a mile broad in places, and bounded by banks of Boulder Clay about 50 feet in height, this thickness of clay having been removed in the formation of the valley. But south of Aldford the broad flat through which the river winds shades off on the west into a scarcely perceptible slope of Boulder Clay. In this case the work of the river has been rather to fill up than to excavate, the gradient not having been sufficient to produce currents of transporting power.

* At the new Gasworks by the side of the railway the following section was met with in December 1880 :—

|  | | | | | feet. |
|---|---|---|---|---|---|
| Sand and Silt | - | - | - | - | 5—6 |
| Boulder Clay | - | - | - | - | 25+ |

The same observation may be made with regard to the alluvium of the Gowy north of Barrow, and that arm of the marshes of the Mersey which forms Hapsford Moor. In both cases depressed areas of Boulder Clay, not excavated by the present rivers, but existing as depressions in consequence of the depth of the rock-surface in the old valley beneath, have received broad deposits of fluviatile mud. The amount of denudation of the drift in such situations has been exceedingly small, and even in more favourable localities has been confined to the planing off a small portion of the surface, as the river, in winding, has undermined the one bank or the other. In no case has the complete removal of the drift from the pre-glacial valley been nearly accomplished.

It is probable that if the Drift had no existence, some portions of the river valleys would form lake-basins. In that of the Dee, for example, a natural barrier of rock exists at Chester, while higher up the channel is wholly in Drift, probably of considerable thickness. So gentle is the gradient in the lower part of this river, that high spring tides flow 12 or 14 miles above Chester. It is known that in many parts of Cheshire (as well as Lancashire) the surface of the rock is below the sea-level, and would therefore, in the absence of the Drift, be overspread by the sea.

There runs from the Dee, near Chester, to the Gowy, near Stoke, a very distinct valley, occupied for its whole length by a strip of marsh, which connects the alluvia of the two rivers. It has been stated that there has been a free passage between the waters of the Dee and the Mersey by this channel within historic times, the evidence adduced being (1) the continuity of the alluvium; (2) the discovery of marine shells of recent species in the valley; (3) the authority of an old map, of the time of John Scot, eighth Earl of Chester, 1232–1237, on which the valley is represented as being occupied by water.

But in the first place, the alluvium, as may be readily seen by comparison with the level of the canal, rises from either end towards the middle, where it reaches a height of 40 feet above the Ordnance Datum. This form can only be due to the action of streams running from a central watershed in opposite directions. Moreover, a submergence sufficient to enable the tide to overflow this barrier would place below the sea-level many of the ancient roads of the district. Secondly, the marine shells are now known to have been obtained from the glacial sand, which crops out in the sides of the valley. Thirdly, in this same map Aldford Brook is also drawn as a continuous stream flowing between the Dee into the Gowy by Tattenhall. It can be asserted with certainty that, though the watershed is low, and may have been swampy, such a continuation of this brook has never existed. It is probable that in both these cases the indefinite position and the swampy nature of the watershed led to the existence of continuous water-channels being assumed.

# PART III.
## ECONOMIC GEOLOGY.

### *Agricultural.*

The edition of this map for the geology of the superficial or Drift deposits shows, the distribution of those formations which underlie and give rise to the soil, and therefore indicates the character of the soil for any locality. The greater part of the area, especially the low-lying clay-covered districts on the east and west side of the Keuper Sandstone escarpment, has been under cultivation from time immemorial, but the less fertile Sandstone and Drift Sand districts of the Alvanley, Delamere, Eddisbury, and Peckforton hills were only reclaimed at the end of the last or the beginning of the present centuries, and are still in part heather-covered, or devoted to growing Scotch Fir, Larch, and stunted Oaks.

The formations may be classified according to their agricultural character as follows :—

SANDY SOIL    -    $\left\{\begin{array}{l}\text{Drift Sand and Gravel.}\\\text{Sandstone (in absence of Drift).}\end{array}\right.$

CLAY SOIL    -    $\left\{\begin{array}{l}\text{Boulder Clay.}\\\text{Keuper Marl.}\\\text{Waterstones.}\end{array}\right.$

*Sandy Soils.*—The contrast between the Drift Sand and the Boulder Clay areas is the more marked from the tendency of the former to occupy higher ground, the rapid drainage from which increases the tendency of the soil to "burn" in the summer. Immediately under the soil, at a depth of 4 to 16 inches, there usually occurs a seam or "pan" of sand, cemented into a solid mass by oxide of iron, and locally known as Fox-bench.

The water circulating in the sand contains a variable proportion of ferrous salts in solution. On exposure to air, such a solution absorbs oxygen and deposits hydrous oxide, a process which may be seen to be taking place in many springs, where they issue into the open air, and which has caused the name " cary " (Lat. *Caries* ?) to be applied to such waters. From the porosity of the soil in the sand areas the oxygen is enabled to circulate in it, and the iron is deposited in a layer beneath the surface. The effect of the Fox-bench is to prevent the penetration of roots. After it has been broken up by deep ploughing, its " growth " may be prevented by draining. Quicklime is said to counteract its evil effects.

The Sandstone area includes that occupied by the Pebble Beds and the Lower Keuper Sandstone. The former rise to the surface in isolated patches, and make naturally a light and fertile soil. The gentle hills formed by this rock from their comparative dryness offered to the early inhabitants of the country more favourable building-sites than the clay-covered plains. The City of Chester and the villages of Aldford, Eccleston, Saighton,

Q 4293.                                                    c

Christleton, Waverton, Thornton, Hapsford, Dunham, Barrow, Tarvin, and Tattenhall are in such situations. The Lower Keuper Sandstones from their purely quartzose character produce a soil more resembling that of the Drift Sand. They form the highest and steepest ground of the district, so that the soil is liable to be washed away as fast as created.

In reclaiming the Drift Sand and Lower Keuper Sandstone areas, it has been found necessary to create an artificial soil by marling.

For this purpose Boulder Clay and the shales of the Waterstones and Keuper Marls have been used with equally good results; the former is known as Clay Marl and the two latter as Slate Clay. The association of the Waterstones with the Lower Keuper Sandstone and their occurrence at high elevations is exceedingly fortunate in this respect. The positions of the principal marl-pits are recorded on the map, and the most favourable spots for opening others may be determined. I am informed by Mr. Harrison, of the New Pale Farm, that the Finney Hill marl-pit is considered one of the best in the neighbourhood. Good marl should effervesce slightly with dilute hydrochloric acid (showing the presence of carbonate of lime) and should readily crumble in frost or in rain after dry weather. The lime may sometimes be seen in crystals lining small cavities in the stone-bands. It occurs in both the red and green beds, but the latter more frequently effervesce with acids.

The following account of marling operations in Delamere Forest* is taken from a paper by Mr. R. Grantham in the Journal of the Royal Agricultural Society, vol. xxv. The forest was planted with oak by the Commissioners of Woods, in pursuance of an Act of Parliament passed in 1812, for the purpose of supplying timber for the navy. The young crop having partly failed in 1856, portions of the land were cleared, marled, and let on farming leases.

The Houndslow Farm, comprising 248 acres, was marled by 29,000 cubic yards taken from the pit in Keuper Marls, figured on p. 30, at a total cost of 1,797*l*. The letting value of the land was thereby increased from 5*s*. to 30*s*. The following is an analysis of an average sample of the marl:—

| | |
|---|---|
| Moisture, organic matter | 4·48 |
| Oxides of iron and alumina | 14·21 |
| Carbonate of lime | 8·65 |
| Magnesia | 1·39 |
| Phosphoric acid | 0·36 |
| Potash and soda | 1·91 |
| Insoluble siliceous matter | 69·00 |

The Long Ridge and Glover's Moss Farm, comprising 800 acres, was marled from a bed of dull red clay,† under the N.E. slope

---

* A variety of sheep occurred on the Forest of Delamere, " black, brown, or grey, " or with spotted faces or legs, and usually small horns; they are not unlike a " diminutive Norfolk. . . . The wool is short and particularly fine. After the " enclosure of the forest they were dispersed through the country."—" The Farming of Cheshire," Journ. Roy. Agric. Soc., vol. v. (1845).

† Boulder Clay.—A. S.

of Eddisbury Hill in years 1861–63. The bed of marl was 12 to 15 feet thick, and rested on a bed of sand and round pebbles. It thinned out on both sides. 87,228 cubic yards were conveyed an average distance of 2 miles and spread on the land at a total cost of 7,890*l.* The letting value of the land increased from 5*s.* to 32½*s.* An analysis gives:—

| | | | | |
|---|---|---|---|---|
| Moisture, organic matter | - | - | - | - 2·40 |
| Oxides of iron and alumina | - | - | - | 16·53 |
| Carbonate of lime | - | - | - | - 5·76 |
| Magnesia | - | - | - | - 2·11 |
| Phosphoric acid | - | - | - | trace. |
| Potash and soda (chlorides) | - | - | - | - 2·01 |
| Carbonic acid | - | - | - | - 7·71 |
| Insoluble siliceous matter | - | - | - | - 63·48 |

The Organs Dale and Primrose Hill Farm of 530 acres was marled from the Waterstones in the forest east of the Heald ; the letting value of the land increased from 7*s.* to 33*s.* per acre. An analysis is as follows :—

| | | | | |
|---|---|---|---|---|
| Moisture, organic matter | - | - | - | - 2·91 |
| Oxides of iron and alumina | - | - | - | - 8·10 |
| Lime | - | - | - | - 2·84 |
| Magnesia | - | - | - | - 2·40 |
| Phosphoric acid | - | - | - | - trace. |
| Potash and soda (as chlorides) | - | - | - | 1·58 |
| Carbonic acid | - | - | - | - 4·33 |
| Insoluble siliceous matter | - | - | - | - 77·84 |

The Old and New Pale Farms were reclaimed at the latter end of the last century and the beginning of this, under leases from the Crown. Both were marled from the Waterstones with a good and lasting effect. The Old Pale has been in part re-marled. In 1846 a railway 100 yards long was laid to marl land at Alvanley which had previously been almost bare of soil and supported only dwarfish heath.* Simmond's Hill was marled in 1876–78 by means of a tramway laid to the marl-pit near Irk Wood (described on p. 11).

Marling, however, is now rarely practised on cultivated land on account of the greater cost of labour, and the increasing use of bone manure. But it is probably the cheapest way of reclaiming heath and creating a soil where it is not naturally supplied. After the lapse of years the marl has been found to collect in a layer a few inches below the subsoil.

*Clay Land.*—Marling was also extensively practised over the Boulder Clay and Keuper Marl area, but has now been supplanted by the use of bones and farmyard manure. The old marl pits, which are very numerous, are still useful as drinking places for cattle. Careful draining is essential to the cultivation of the Boulder Clay. The greater part of the area is under pasture, the labour of ploughing and breaking up the clods in the stiff clay being very great. Further information on the mode of farming in Cheshire may be found in the Journal of the Royal Agricultural Society, vols. v., xix., xxv., vi. (2nd series).

---

* " Manuring Grass Lands."—J. Dixon, Journ. Roy. Agric. Soc., vol. xix. (1858).

## Building Materials, &c.

The *Keuper Marls* have been manufactured into bricks at Cuddington Waste.

The quarries and marl-pits in the *Waterstones* and the building-stones of the *Lower Keuper Sandstone* have been described previously. Some of the soft beds in the latter are occasionally used as building-sand. The *Upper Mottled Sandstone* has been used for sanding the floors of cottages and gardening purposes. The white and softer beds only are used; they have been excavated in irregular caves and passages in the side of the hill at Beeston and at the north-east angle of Helsby Hill, where the roof of the cave is formed of Keuper Sandstone, and at other places near Frodsham (Explanation 80 N.W.).

The quarries in the *Pebble Beds* have been already described.

The *Drift Sand* is used everywhere for building purposes, and the beds of gravel sifted for road metal. The latter are valuable in such districts as Delamere Forest, where metal is scarce. They occur very irregularly, but as a general rule are more common near the hills than in the plains.

The principal sand and gravel pits are marked upon the map.

The *Boulder Clay* is the chief source of bricks in the district. It is turned over to a depth of 6 feet in the autumn, and left to temper during the winter; the boulders are picked out by hand, and used as paving or are broken up for road-metal. There are numerous brick-pits in the suburbs of Chester, chiefly in the neighbourhood of Boughton and Bishop's Fields. The clay is often found to be of good quality for brick-making close to its margin, where it is thinning off upon a rising slope either of rock or sand. Large quantities of bricks were supplied to the city formerly from the hollow now occupied by the cemetery, and those used in building the bridge of the Cheshire Lines across the Holyhead Railway were made on the spot. In both cases the pits were situated close to the margin of the clay, so that the rock was laid bare in the excavations.

### Water Supply.

Of the total area of 168 square miles in this Quarter-sheet about 14 per cent. are occupied by formations permeable to rainfall, the Waterstones and the Drift Sand where underlain by an impervious formation not being included. The villages situated on the rock draw their supply from shallow wells extending to a short distance below the water-level; but the greater part of the district is supplied by wells sunk through the Boulder Clay into the Drift Sand.

In the low-lying areas it is found that the water rises rapidly in a well as soon as the Boulder Clay is penetrated, and a sufficient supply being thus provided, further sinking becomes unnecessary. In such localities as Cotton Abbots, Tattenhall (p. 25), where the water is necessarily derived from a distance owing to the extent of clay-covered ground, it is probably as good as the deep well water from the rock, and can hardly be classed under the head of

" land-springs "; but it would probably be found to be inadequate for the supply of towns. In the case of villages built on Drift Sand, a local supply from shallow wells must be regarded as *highly suspicious.* The rainfall charged with surface impurities is at once absorbed by the porous subsoil, and a large area may be contaminated by the impurity of a single spot.

The water-bearing capability of the rock has not yet been tested. By the experience of similar conditions in Lancashire, it may be presumed that in the Bunter occupying the western portion of the sheet, there exists a large supply practically untouched hitherto. The strata being practically of equal porosity throughout, the water may be considered to lie in a sheet, the surface of which rises from the sea-level on the coast, towards the higher ground inland. While the level in the Pebble Bed hills is at a depth below the surface proportionate to the height and steepness of the hill, in the intervening low ground it can be little below, and probably is often above, the surface, as is the case in the low-lying areas of Cronton and Netherlea (80 N.W.). The rock underlying such areas being of an exceedingly porous nature, the supply would probably be large under favourable local conditions.

As to the quality of the water likely to be met with, there is at present little information. From the experience of other localities, it is pretty certain to be palatable and wholesome, though possibly harder than is convenient for household purposes.

In 1876 a well was sunk at the Duke of Westminster's Waterworks on the banks of the river Dee for the purpose of supplying Eaton Hall.* The well is 40 feet deep with a bore-hole for a further 307 feet in the Pebble Beds of the Bunter. It is situated at 20 yards from the river, and 14 feet above it. The water in the well stands at 8 feet from the surface, or 6 feet above the water in the river. The supply is 37½ cubic feet per minute.

An analysis of the water by Mr. Ogston is as follows:—

| One gallon contains, solid residue | | | | | 63·20 grains. |
|---|---|---|---|---|---|
| Volatile matter | - | - | - | - | 2·00 |
| Chlorine | - | - | - | - | 15·20 |
| Sulphuric Acid | - | - | - | - | 9·36 |
| Lime | - | - | - | - | 11·20 |
| Magnesia | - | - | - | - | 5·98 |
| Oxide of Iron | - | - | - | - | ·60 |
| Siliceous matter | - | - | - | - | ·38 |
| Alkalies and Carbonic Acid | - | - | - | - | 15·34 |
| Nitric Acid | - | - | - | - | 2·64 |
| | | | | | 63·20 |

| Total Hardness | - | - | - | - | - | 20°·5 |
|---|---|---|---|---|---|---|
| Permanent Hardness | - | - | - | - | - | 15°·0 |
| Free ammonia per gallon ·001 ; albuminoid ammonia | | | | | - | ·0021 |

As the water was found to be unfit to be used for washing or in boilers a plug was inserted in the bore-hole and the lower 150 feet stopped off. The water was then pumped down to 25 feet 10 inches from the surface in one hour, the engine raising

---

* The well has since been abandoned,

250 gallons per minute; it rose to its former level in eight hours. It was found to be less hard in the upper than in the lower portion of the bore-hole. An overflow pipe was put in at 7 feet 9 inches from the surface (the water at that time standing at 5 feet from the surface) and carried off a strong continuous stream. The water after being pumped down to 25 feet 10 inches at the same rate as previously, rose again to the overflow pipe in 1 hour 20 minutes, showing that the entry of water into the well, whether from fissures or pores in the rock, is very much retarded as the level in the well approaches the water-level in the rock, probably owing to the diminution of pressure.

The water which is now overflowing was kept down in the rock at this point by the 20 feet of overlying Boulder Clay; but there is probably a natural overflow at the nearest point at which the rock rises to the surface at the level of the river. There are springs possibly of this nature at the edge of the alluvial marsh west of Aldford.

Good water is obtained from a shallow public well in Pebble Beds in Eccleston.

*Chester.*—In a well at Seller's Brewery, Seller Street, a good spring was struck at 48 feet from the surface in soft, white rock; the water is not very hard. Huxley's Brewery, King Street, is supplied from an old well in Pebble Beds, 70 to 80 feet deep. It is pumped dry in four to five hours, and the water is said to be of good quality and not very hard.

In Queen's Park* in a well 36 feet deep in Pebble Beds, the water stands at more than 30 feet from the surface; it is not hard and always clear, but deposits iron if left to stand in a vessel.

The City of Chester is supplied with water from the river Dee, received at a distance of 1,350 yards above the pumping station and filtered. In 1849 † a shaft was sunk in Red Sandstone (Pebble Beds) at the pumping-station on the banks of the river; a good spring was met with, but the water was very hard. In the hopes of getting river-water naturally filtered by passage through sandstone, a tunnel was driven under the river at a presumed depth of 19 feet below the bed of the river for a distance of eight yards; this yielding no water, a second was driven within 6 feet of the bed of the river for 6 yards with a like result; tunnels were then driven at the lower level both up and down the river and beneath its bed, that on the north extending for 38 yards. The quantity of water was not increased by the additional tunnelling and the undertaking was abandoned.

*Springs.*—The following are the principal springs in the district:—

| | | |
|---|---|---|
| W. of Aldford | from | Pebble Beds. |
| Horsley Bath (abundant) | „ | Upper Mottled Sandstone. |
| Willington Corner | „ | Lower Keuper Sandstone. |
| Whistlebitch Well | „ | „    „ |

---

* From Dr. Stolterfoth.
† From Mr. W. Brown.

| | | | |
|---|---|---|---|
| Near Wood Lane, Utkinto | from | Lower Keuper Sandstone. | |
| Eaton, near Tarporley | ,, | Waterstones. | |
| Eddisbury Hill, N.E. side | ,, | ,, | |
| Kingswood (abundant) | ,, | ,, | |
| Crewe railway, near Christleton | ,, | Junction of Boulder Clay and Sand. | |
| Boughton, the Barrel Well* | ,, | ,, | ,, |
| Wimbold Trafford | ,, | ,, | ,, |
| Kingsley | ,, | ,, | ,, |
| Higher Heys | ,, | ,, | ,, |
| Stanney Brook | ,, | ,, | ,, |
| Ashton | ,, | ,, | ,, |
| South of Kelsall | ,, | ,, | ,, |
| Burton | ,, | ,, • | ,, |
| Clotton | ,, | ,, | ,, |
| Oulton | ,, | ,, | ,, |
| Rushton | ,, | ,, | ,, |
| Higher Hall | ,, | Drift Sand. | |
| Shaw Brook (abundant) | ,, | ,, | |
| Bunbury | ,, | ,, | |
| Burton | ,, | ,, | |

There is said to be a mineral spring at Spurstow † in a field forming part of the rising ground towards the Peckforton Hills. The water is at first opaline, but becomes clear on standing. It contains 190 grains of dried solids to the gallon, of which 50 are purgative salts with a little chloride of lime, 120 sulphur, and 20 carbonate of lime.

Some weak brine springs are said to exist near Crewood Green in the township of Kingsley.†

Brine is said to have been worked to a small extent near Aldersey. "The springs are met with near Aldersey Hall; they " rise from drift-clay and sand which occupy the centre of the " valley."† The brine is probably derived from the underlying rock, the Lower Mottled Sandstone. Though rock-salt has never been found in the Bunter, salt-springs are not uncommon. In a borehole at Warrington, sunk through the Upper Mottled Sandstone into the Pebble Beds, strong brine was met with (Memoir on 80 N.W. Ed. 3), and in the Duke of Westminster's well, near Eaton Hall, described on p. 37, a large proportion of Chlorides was proved in the analysis. I am also informed by Mr. De Rance that salt-springs rise from the middle division of the Upper Mottled Sandstone at the Robin Hood Well and Salt Pit Houses near Charnock Richard, between Wigan and Preston. Coal-measures are frequently found to contain brine, and, on the supposition of their underlying the Trias at Aldersey and Warrington at no great depth, may have supplied the salt-springs found at these places. Mr. Beckett informs me that bad water is met with under the Boulder Clay at Poulton, Pulford, at the Rake, and at Cheaveley. At the last-named there were 204 grains of solid matter to the gallon.

---

* A Roman altar, bearing the inscription *Nymphis et Fontibus. Leg.* xx. v. v., was found near this spring. History of the City of Chester, by Joseph Hemingway. Chester, 1831.

† Ormerod, Quart. Journ. Geol. Soc., vol. iv., p. 262.

# APPENDIX.

## LIST OF WORKS ON THE GEOLOGY, MINERALOGY, AND PALÆONTOLOGY OF CHESHIRE,

### By WILLIAM WHITAKER, B.A., F.G.S.

Reprinted, with many Additions, from Proc. Liverpool Geol. Soc.,
pp. 127–147.

(1876.)—For many of the Additions I have to thank Messrs. DALTON, DE RANCE, and STRAHAN.—W.W.

## GEOLOGICAL SURVEY PUBLICATIONS.

*Maps, scale an inch to a mile, price 3s. each quarter-sheet.*

SHEET 73, S.W. (small piece on N.). By *Prof. E. Hull*, 1855.

SHEET 73, N.W. (N.E. part : Malpas). By *A. R. Selwyn* and *Prof. E. Hull*, 1855.

SHEET 73, N.E. half : Crewe, Nantwich). By *W. W. Smythe, A. R. Selwyn*, and *Prof. E. Hull*, 1857.

SHEET 79, S.E. (N.E. corner). By *Prof. E. Hull*, 1850. Additions in 1855.

SHEET 79, N.E. (greater part : Birkenhead, Neston). By *Prof. E. Hull*, 1850. New Edition in 1855.

SHEET 80, S.W. (Chester, Delamere Forest). By *Prof. E. Hull*, 1855. New Edition by *A. Strahan* engraving.

SHEET 80, S.E. (Northwich, Middlewich). By *Prof. E. Hull*, 1858.

SHEET 80, N.W. (S. part : Frodsham, Runcorn). By *Prof. E. Hull*, 1859. New Edition by *A. Strahan* engraving.

SHEET 80, N.E. (all but N. edge : Altrincham and Knutsford). By *Prof. E. Hull*, 1861.

SHEET 81, S.W. (Macclesfield and Congleton). By *Prof. A. H. Green*, in 1864.

SHEET 81, N.W. (greater part : Stockport). By *Prof. E. Hull*, in 1864.

SHEET 88, S.W. (E. part : Staleybridge). By *Prof. E. Hull*, in 1863.

*Horizontal Sections*, scale 6 inches to a mile. Price 5s. a sheet. Each accompanied by an Explanation, price 2d.

SHEET 18 (part), Section of the Red Marl plain of Chester, across the Lower Carboniferous Rocks of North Staffordshire, &c. By *Prof. A. H. Green*. Edition 2, 1866.

SHEET 41, From the South West to North East across Lower Lias. New Red Sandstone . . . . through Norton, Whitmore Heath. By *Prof. B. Hull*, and *A. H. Green*, 1857.

SHEET 43 (part), Section across . . . . . . the New Red Sandstone of Chester and Delamere Forest. By *Prof. E. Hull*, 1858. New Edition by *A. Strahan* preparing.

SHEET 44, No. 1 (part), Section . . . . across . . . . the New Red Sandstone of Holt and the Peckforton Hills. By *Prof. E. Hull*, 1850.

SHEET 64, Section from Northwich on the South-west to Marsden, Yorkshire, on the North-East, &c. By *Prof. E. Hull*, 1864.

SHEET 65, No. 2 (part), From Bowdon, Cheshire, on the West, to Westend Moor, Derbyshire, on the East, &c. By *Profs. E. Hull* and *A. H. Green*, 1865.

SHEET 67, from Arley Hall, Cheshire, on the South-East, to Windle Moss, Lancashire, on the North-West, &c. By *Prof. E. Hull*, 1865.

SHEET 68, Section from Little Eye Island in the Estuary of the Dee on the West, to Horwich Moor, Lancashire, on the East, by Birkenhead, Liverpool . . . . . By *Prof. E. Hull*, 1865.

SHEET 69, Section from the New Red Sandstone of Doghill Green, Cheshire, across the Poynton Coal-field . . . . . the Peak, Derbyshire, . . . . . By *Prof. E. Hull* and *A. H. Green*, 1867. Edition 2, 1872.

SHEET 70 (part), from Swan Bank, near Alderley Edge, across the New Red Sandstone Plain of Cheshire, &c. By *Prof. E. Hull*, 1867.

*Vertical Section*, scale 40 feet to an inch, price 3s. 6d.

SHEET 34. Vertical Sections of the Lancashire and Cheshire Coal-fields. By *Prof. E. Hull*, 1870.

*Memoirs.* 8vo. London.

The Geology of the Country around Prescot, Lancashire (refers partly to Cheshire). By *Prof. E. Hull*, 1860. Edition 2, 1865. Edition 3, by *A. Strahan*, in press.

The Geology of the Country around Altrincham, Cheshire. By *Prof. E. Hull*, 1861. Price 1s.

The Geology of the Country around Oldham, including Manchester and its Suburbs. By *Prof. E. Hull*, 1864. Price 2s.

The Geology of the Country round Stockport, Macclesfield, Congleton, and Leek. By *Profs. E. Hull* and *A. H. Green*, 1866. Price 4s.

The Triassic and Permian Rocks of the Midland Counties of England. By *Prof. E. Hull*, 1869. Price 5s.

The Superficial Geology of the Country adjoining the Coasts of South-west Lancashire. By *C. E. De Rance*, 1877. Price 10s. 6d. (Refers to Cheshire.)

The Geology of the Yorkshire Coalfield. By *A. H. Green* (and others), 1878. Price 2l. 2s. (Refers to the Etherow Valley, see Index.)

---

## LIST OF BOOKS, PAPERS, &c.

### 1630.

*Leigh, C.* Natural History of Lancashire, Cheshire, and the Peak in Derbyshire, &c. *Fol. Oxon.* ? another Edition in 1700.

### 1669.

*Jackson, Dr. W.* Some Inquiries concerning the Salt-Springs and the Way of Salt-making at Nantwich in Cheshire. Answered. *Phil. Trans.*, vol. iv., No. 53, p. 1060.

### 1670.

*Martindale, A.* Extracts of two letters from Rotherton in Cheshire, concerning the Discovery of a Rock of Natural Salt in that County. *Phil. Trans.*, vol. v., No. 66, p. 2015.

### 1684.

*Lister, Dr. M.* Certain Observations of the Midland Salt-Springs of Worcester-shire, Stafford-shire, and Cheshire. *Phil. Trans.*, vol. xiv., No. 156, p. 489.

### 1691.

*Ray, J.* Collection of English Words not generally used, with Account of the Making of Salt at Nantwych in Cheshire, &c.   Ed. 2, 12mo.

### 1740.

*Short, Dr. T.* An Essay Towards A Natural, Experimental, and Medicinal History of the Principle Mineral Waters of . . . . . . . Cheshire, &c. 4to. *Sheffield.*

### 1781.

*Jars, M. G.* Voyages Metallurgiques, tome 3.   (Cheshire, p. 332.)   4to. *Paris.*

### 1795.

*Aikin, J.* Description of the Country from thirty to forty miles round Manchester.  4to.

### 1809.

*Farey, J.* Observations on a late Paper by Dr. W. Richardson, respecting the basaltic District in the North of Ireland, and on the Geological Facts thence deducible; in conjunction with others observable in Derbyshire and other English Counties . . . . . &c.   *Phil. Mag.*, vol. xxxiii., p. 257.

### 1810.

*Farey, J.* A List of about Five Hundred Collieries in and near to Derbyshire.   *Phil. Mag.*, vol. xxxv., p. 431.

### 1811.

*Bakewell, R.* [Description of Alderley Edge, in Cheshire.]   *Monthly Mag.*, vol. xxxi., No. 209, p. 7.
*Farey, J.* A List of about 280 Mines of Lead,—some with Zinc, Manganese, Copper, Iron, Fluor, Barytes, &c., in and near to Derbyshire.   *Phil. Mag.*, vol. xxxvii., p. 106.
*Holland, H.* A Sketch of the Natural History of the Cheshire Rock-Salt District.   *Trans. Geol. Soc.*, vol. i., p. 38.

### 1813.

*Holland, H.* General View of the Agriculture of Cheshire.   8vo.  *London.* (Nature and Origin of Marl, by J. F. Stanley, pp. 348-354.   Comparative View or the Theories relative to the origin of Rock Salt, pp. 355-369.)

### 1817.

*Aikin, A.* Notice on a Green waxy Substance found in the alluvial soil near Stockport, in Cheshire.   *Trans. Geol. Soc.*, vol. iv., p. 445.
*Trail, Dr.* Notice of some magnetic iron sand, Cheshire.   *Ibid.*, p. 447.

### 1824.

*Anon.* Fossil Elephant's Tooth found in Cheshire.   *Edin. Phil. Journ.*, vol. xi., p. 417.

### 1825.

*Sedgwick, Rev. Prof. A.* On the Origin of Alluvial and Diluvial Formations. (Cheshire, p. 23.)   *Ann. of Phil.*, Ser. 2, vol. x., p. 18.

### 1826.

*Hibbert, Dr. S.* On some Remarkable Concretions which are found in the Sandstone of Kerridge in Cheshire.   *Edin. Journ. of Sci.*, vol. iv., p. 138.

### 1828.

*Stevenson, R.* Remarks upon the Wasting Effects of the Sea on the shore of Cheshire, between the rivers Mersey and Dee.   *Edin. New Phil. Journ.*, vol. iv., p. 386.

1829.

*Brayley, E. W.* On the Existence of Salts of Potash in Brine-Springs and in Rock Salt. *Phil. Mag.*, Ser. 2, vol. v., p. 411.
*Nicol, W.* On the cavities containing Fluids in Rock-Salt. *Edin. New Phil. Journ.*, vol. vii., p. 111.

1830.

*Daubeny, Dr. C.* Memoir on the occurrence of Iodine and Bromine in certain Mineral Waters of South Britain. (Cheshire, pp. 229, 230.) *Phil. Trans.*, vol. cxx., p. 223.
*Oeynhausen, C. von, and H. von Dechen.* [On Rock Salt in England.] *Karsten's Archiv.* 1, i., p. 56.

1832.

*Hall, E.* A Mineralogical and Geological Map of the Coal Field of Lancashire, with parts of . . . . . Cheshire . . . . . (an inch to a mile). *Manchester.* 2 other Editions.

1833.

*Trimmer, J.* Discovery of marine shells of existing species on the left bank of the river Mersey, and above the level of high-water mark. *Proc. Geol. Soc.*, vol. i., p. 419.

1835.

*Egerton, Sir P. G.* On a Bed of Gravel containing Marine Shells of recent species, at "The Willington," in Cheshire. *Proc. Geol. Soc.*, vol. ii., p. 189.
*Murchison [Sir] R. I.* On an outlying basin of Lias on the borders of Salop and Cheshire, &c. *Ibid.*, p. 114.
——— The Gravel and Alluvia of S. Wales and Siluria as distinguished from a northern drift covering Lancashire, Cheshire, &c. . . . . *Ibid.*, p. 230.

1836.

*Egerton, Sir P. De M. G.* A notice on the occurrence of marine shells in a bed of gravel at Norley Bank, Cheshire.—*Proc. Geol. Soc.*, vol. ii., p. 415.
*Hall, E.* Introduction to the Mineral and Geological Map of the Coalfield of Lancashire with a part of . . . . . Cheshire. Pp. iv. 28. 8vo. *Manchester.*

1837.

*Anon.* Cheshire. "Geological Character." *Penny Cyclopædia*, vol. vii., p. 43. (*Fol. Lond.*)
*Tooke, A. W.* The Mineral Topography of Great Britain. (Cheshire, p. 40.) *Mining Review*, No. 9, p. 39.

1838.

*Anon.* (Liverpool Natural History Society.) An account of Footsteps of the Chirotherium, and other unknown animals lately discovered in the quarries of Storeton Hill, in the peninsula of Wirrall, between the Mersey and the Dec. . . . . . illustrated with drawings by J. CUNNINGHAM. *Proc. Geol. Soc.*, vol. iii., p. 12.
*Denham, Capt.* On the Tidal Capacity of the Mersey Estuary—the Proportion of Silt held in solution . . . . . the Excess of Deposit upon each Reflux, and the consequent Effect . . . . . *Rep. Brit. Assoc. for* 1837, *Trans. of Sections*, p. 85.
*Egerton, Sir P. De M. G.* On two Casts in Sandstone of the impressions of the Hind Foot of a gigantic Chirotherium, from the New Red Sandstone of Cheshire. *Proc. Geol. Soc.*, vol. iii., p. 14.
*Trimmer, J.* On the Diluvial or Northern Drift of the Eastern and Western Sides of the Cambrian Chain, and on its Connexion with a similar Deposit on the Eastern Side of Ireland. . . . . *Journ. Geol. Soc. Dublin*, vol. i., part 4, pp. 286, 335.

1839.

*Cunningham, J.* An Account of Impressions and Casts of Drops of Rain, discovered in the Quarries of Storeton Hill, Cheshire. *Proc. Geol. Soc.*, vol. iii., p. 99.

*Egerton, Sir P. De M. G., and J. Taylor.* Letters on a slab of sandstone, exhibiting footmarks, and supposed to be from the Kelsall quarry. *Ibid.*, p. 100.

*Grant, Prof.* [R. E.] Footmarks of Chirotherium at Stourton Hill. *Mag. Nat. Hist.*, Ser. 2, vol. iii., p. 43.

1840.

*Buckland, Rev. Prof. W.* (On fossil impressions of rain, and ripple marks . . . . and fossil footsteps of Cheirotherium and other unknown animals recently discovered on strata of the new red sandstone formation in the counties of Cheshire, Salop, and Warwick.) *Proc. Ashmolean Soc., Oxon.*, No. xvi., p. 5.

*Hodgkinson, E.* On the Temperature of the Earth in the deep Mines of Lancashire and Cheshire. *Rep. Brit. Assoc. for 1839, Trans. of Sections*, p. 19.

*Yates, J. B.* On the rapid changes which take place at the Entrance of the river Mersey, and the means adopted for establishing an easy access to Vessels resorting thereto. *Ibid.*, p. 77.

1841.

*Binney, E. W.* Sketch of the Geology of Manchester and its Vicinity. *Trans. Manchester Geol. Soc.*, vol. i., p. 35.

——— Observations on the Lancashire and Cheshire Coal Field. *Ibid.*, p. 67.

*Bullfinch, —.* Report on the Casts of Fossil Footsteps from Cheshire, England. *Proc. Boston Nat. Hist. Soc.*, p. 45.

*Hodgkinson, E.* On the Temperature of the Earth in the deep Mines in the neighbourhood of Manchester. *Rep. Brit. Assoc. for 1840, Trans. of Sections*, p. 17.

1842.

*Heywood, J.* Remarks on the Coal District of South Lancashire. (Refers to Cheshire.) *Mem. Lit. Phil. Soc., Manchester*, Ser. 2, vol. vi., p. 426.

*Ormerod, —.* On the Salt of Cheshire. *Geologist*, p. 150, and *Trans. Manchester Geol. Soc.*, vol. viii., p. 25. (1869.)

1843.

*Binney. E. W.* Notes on the Lancashire and Cheshire Drift. *Geologist*, p. 112, and *Trans. Manchester Geol. Soc.*, vol. viii., p. 30. (1869.)

——— On the Great Lancashire and Cheshire Coal Fields. [Brit. Assoc.] Reprinted from *Annals of Philosophical Discovery and Monthly Reporter of the Progress of Practical Science.* 10 pp. 8vo. *Manchester*.

*Buckland, Rev. Dr. W.* On Recent and Fossil Semi-circular Cavities caused by air-bubbles on the surface of soft clay, and resembling impressions of raindrops. *Rep. Brit. Assoc. for 1842, Trans. of Sections*, p. 57.

*Hawkshaw, —.* Notice of the Fossil Footsteps in the New Red Sandstone Quarry at Lymm, in Cheshire. *Ibid.*, p. 56.

*Ormerod, G. W.* On the Geology of Central Cheshire. *Geologist*, p. 1, 2 and *Trans. Manchester Geol. Soc.*, vol. viii., p. 25. (1869.)

1845.

*Hume, Rev. Dr. A.* An Account of a recent Visit by several Members of the Society, to the Submarine Forest at Leasowe. *Proc. Lit. Phil. Soc. Liverpool*, vol. i., p. 97.

1846.

*Binney E. W.* Description of the Dukinfield Sigillaria. *Quart. Journ. Geol. Soc.*, vol. ii., p. 390.

——— On the Relation of the New Red Sandstone to the Carboniferous Strata in Lancashire and Cheshire. *Quart. Journ. Geol. Soc.*, vol. ii., p. 12.

*Black, Dr. J.* Observations on a Slab of New Red Sandstone from the Quarries at Weston, near Runcorn, Cheshire, containing the Impressions of Footsteps and other markings. *Ibid.*, p. 65.

*Hume, Dr.* [*A.*] Notes of an Oral Lecture, delivered on Geological Subjects during a recent Excursion to Stourton. *Proc. Lit. Phil. Soc., Liverpool,* No. ii., p. 52.

## 1847.

*Cunningham* [*J.*] On the Geological Conformation of the Neighbourhood of Liverpool, as respects the Supply of Water. *Proc. Lit. Phil. Soc., Liverpool,* No. iii., p. 58.

*Ormerod, G. W.* On the Extent of the Northwich Salt-field. *Rep. Brit. Assoc. for 1846, Trans. of Sections,* p. 62.

## 1848.

*Binney, E. W.* On the Origin of Coal. *Mem. Lit. Phil. Soc., Manchester,* Ser. 2, vol. ix., p. 148.

———— Sketch of the Drift Deposits of Manchester and its Neighbourhood. *Ibid.,* p. 195.

*Chambers, R.* Ancient Sea-Margins, as Memorials of Changes in the Relative Level of Sea and Land. (Cheshire, pp. 223–228.) 8vo. *Edin. and Lond.*

*Cunningham, J.* Description of Plates (Impressions on Sandstone, &c., Stourton). *Proc. Lit. Phil. Soc., Liverpool,* No. iv., p. 127.

*Ormerod, G. W.* Outline of the principal Geological features of the Salt-field of Cheshire and the adjoining districts. *Quart. Journ. Geol. Soc.,* vol. iv., p. 262.

## 1849.

*Hume, Rev. A.* Notice of certain Mineral Springs at Leasowe. *Proc. Hist. Soc. Lanc. Chesh.,* Sess. 1, p. 105.

*Picton, J. A.* The Changes of Sea Levels on the West Coast of England during the Historic Period. (Abstract.) *Proc. Lit. Phil. Soc., Liverpool,* No. v., p. 113.

*Smith, J. P. G.* [Section of the Bidston Marsh.] *Ibid.,* p. 169, plate.

## 1850.

*Danson, J.* Analysis of the Leasowe Water. *Proc. Hist. Soc. Lanc. Chesh.,* Sess. ii., p. 260.

*Harkness,* [*Prof.*] *R.* Notice of a Tridactylous Footmark from the Bunter Sandstone of Weston Point, Cheshire. *Ann. Nat. Hist.,* Ser. 2, vol. vi., p. 440.

## 1851.

*Dickinson, J.* On the Physical Geography of Liverpool and Wirral. *Appendix* (The Flora of Liverpool) to *Proc. Lit. Phil. Soc., Liverpool,* No. vi.

*Trimmer, J.* On the Erratic Tertiaries bordering the Penine Chain, between Congleton and Macclesfield; and on the Scratched Detritus of the Till. *Quart. Journ. Geol. Soc.,* vol. vii., p. 201.

## 1852.

*Binney, E. W.* Notes on the Drift Deposits found near Blackpool. *Mem. Lit. Phil. Soc., Manchester,* Ser. 2, vol. x., p. 121.

## 1853.

*Rawlinson, R.* On Foot-tracks found in the New Red Sandstone at Lymm, Cheshire. *Quart. Journ. Geol. Soc.,* vol. ix., p. 37.

*Trimmer, J.* On the Erratic Tertiaries bordering the Penine Chain. Part 2. *Ibid.,* p. 352.

## 1855.

*Binney, E. W.* On the Permian Beds of the North-West of England. *Mem. Lit. Phil. Soc., Manchester,* Ser. 2, vol. xii., p. 209.

## 1856.

*Binney, E. W.* On some Footmarks in the Millstone Grit of Tintwistle, Cheshire. *Quart. Journ. Geol. Soc.,* vol. xii., p. 350.

*Boult, J.* On Some of the Recorded Changes in the Liverpool Bay, previous to the Year 1800. (Abstract.) *Trans. Hist. Soc. Lanc. Chesh.,* vol. viii., pp. 253, 254.

*Morton, G. H.* On the Sub-divisions of the New Red Sandstone between the River Dee and the Up-throw of the Coal Measures east of Liverpool. *Proc. Lit. Phil. Soc., Liverpool,* No. x., p. 68.

## 1857.

*Northcote, A. B.* On the Brine-springs of Cheshire. *Phil. Mag.,* Ser. 4, vol. xiv., p. 457.

*Rennie, G., J. Boult, and A. Henderson.* Report from the Committee . . . to investigate and report upon the effects produced upon the Channels of the Mersey by the alterations which within the last fifty years have been made in its Banks. *Rep. Brit. Assoc. for* 1856, p. 1.

## 1858.

*Higgs, S.* Notice of the Copper Mines of Alderley Edge, Cheshire. *Trans. Roy. Geol. Soc., Cornwall,* vol. vii., p. 325.

## 1859.

*Atkinson, J.* (On a curiously-shaped Fossil found in the Upper New Red Sandstone, near Runcorn.) *Proc. Lit. Phil. Soc., Manchester,* vol. i., p. 164.

## 1860.

*Atkinson. J.* On the Drift Deposits in the Neighbourhood of Thelwall, Cheshire. *Trans. Manchester Geol. Soc.,* vol. ii., Part 6, p. 63.

*Binney, E. W.* Observations on the Fossil Shells of the Lower Coal Measures. *Ibid.,* Part 7, p. 72.

*Henderson, W.* Economic Treatment of Poor Copper and other Ores [refers to the Alderley Edge deposits]. *Mining Journal,* vol. xxx., pp. 686, 690.

*Hull, E.* On the New Subdivisions of the Triassic Rocks of the Central Counties. *Trans. Manchester. Geol. Soc.,* vol. ii., Part 3, p. 22.

*Morton, G. H.* [printed "Hillotson."] Slickensides. *Geologist,* vol. iii., p. 37.

———— Evidences of Ancient Ice-action near Liverpool, and on Pleistocene Deposits near Liverpool. *Ibid.,* pp. 197, 349. See also p. 277.

## 1861.

*Mitchener, J. H.* On a New Red Sandstone Quarry at Stourton in Cheshire. *Proc. Geol. Assoc.,* vol. 1., p. 75.

*Morton, G. H.* On the Basement Bed of the Keuper Formation in Wirral and the Southwest of Lancashire. *Proc. Liverpool Geol. Soc.,* Sessions 1 and 2, p. 4.

———— On the Pleistocene Deposits of the Districts around Liverpool. *Ibid.,* p. 12. [Brit. Assoc.]

*Sainter, J. D.* A Salt Spring in a Coal Mine (at Dukinfield). *Geologist,* vol. iv., p. 398.

## 1862.

*Binney, E. W.* [On the Crystal of Selenite found in the till in North Cheshire.] *Trans. Manchester Geol. Soc.,* vol. iii., p. 8.

———— On "Jelly Peat"—found at Churchtown near Southport. *Ibid.,* p. 19.

*Fairbairn, Sir W.* Remarks on the Temperature of the Earth's Crust as exhibited by Thermometrical Returns obtained during the sinking of the Deep Mine at Dukinfield. *Rep. Brit. Assoc. for* 1861, *Sections,* p. 53.

*Hull, E.* Marine Fossils at Dukinfield (in Coal). *Trans. Manchester Geol. Soc.,* vol. iii., p. 348.

*Hull, E.* The Lancashire and Cheshire Coal Fields. *Mining and Smelting Mag.*, vol. i., p. 85.

*Jones, Prof. T. R.* A Monograph of the Fossil Estheriæ. (Cheshire, pp. 64, &c.) *Palæontograph Soc.*

*Taylor, J.* On Pleistocene Deposits on the Stockport and Woodley Railway. *Trans. Manchester Geol. Soc.*, vol. iii., p. 147.

———— On the Geology of the Railway between Hyde and Marple. *Trans. Manchester Geol. Soc.*, vol. iii., p. 296.

1863.

*Binney, E. W.* Section of the Drift near Rainford. *Geol. Mag.*, vol. vi., p. 307. See also p. 308.

*Morton, G. H.* On the Surface Markings near Liverpool, supposed to have been caused by Ice. *Proc. Liverpool Geol. Soc.*, Session 3, p. 9.

———— On the Thickness of the Bunter and Keuper Formations around Liverpool. *Ibid*, p. 15.

———— Report of the Society's Field Meetings at Storeton and Leasowe. *Ibid*, Session 4, p. 5.

———— Description of the Footprints of Cheirotherium and Equisetum, found at Storeton, Cheshire. *Ibid*, p. 17.

———— The Geology of the Country around Liverpool. 8vo. *Liverpool.*

*Platt, —* [Records of Borings at Middlesborough, Bradford, and Wirrall.] *Proc. Lit. and Phil. Soc., Manchester*, vol. iii., No. 3, Session 1863-4, p. 133.

*Sainter, J. D.* Bones at Macclesfield. *Geologist*, vol. vi., pp. 185-187.

*Tate, A. N.* On the Composition of Black Sandstone occurring in the Trias around Liverpool. *Proc. Liverpool Geol. Soc.*, Session 4, p. 16.

1864.

*Barr, W. R.* On the Quaternary Deposits of the Valley of the Mersey, near Stockport. *Trans. Manchester Geol. Soc.*, vol. iv., (No. 15), p. 335.

*Binney, E. W.* A few remarks on the Lancashire and Cheshire Drift. *Proc. Lit. Phil. Soc., Manchester*, vol. iii., No. 10, (Session 1863-4), p. 214, (and *Geologist*, vol. vii., p. 140).

*Cust, Lt.-Gen. Sir E.* The Prehistoric Man of Cheshire : or Some Account of a Human Skeleton found under the Leasowe Shore in Wirral. *Trans. Hist. Soc. Lanc. Chesh., N. Ser.*, vol. iv., pp. 193, 249.

*Darbishire, R. D.* Notes on Marine Shells found in Stratified Drift at Macclesfield. *Ibid*, vol. iv., No. 5, (Session 1864-5), p. 41.

*Eskrigge, R. A.* On the Lias of Cheshire and Shropshire. *Trans. Manchester Geol. Soc.*, vol. iv., (No. 14), p. 318.

*Grantham, R. B.* A Description of the Works for Reclaiming and Marling parts of the late Forest of Delamere, in the County of Cheshire. *Journ. Roy. Agric. Soc.*, vol. xxv., p. 369.

*Greenwell, G. C.* On the Copper Sandstone of Alderley, Cheshire. *Trans. S. Wales Inst. Eng.*, vol. iv., p. 44.

*Hull, (Prof.) E.* On the Occurrence of Glacial Striations on the Surface of Bidston Hill, near Birkenhead. *Trans. Manchester Geol. Soc.*, vol. iv., p. 288.

———— On the New Red Sandstone and Permian Formations, as Sources of Water-supply for Towns. *Mem. Lit. Phil. Soc., Manchester*, Ser. 3, vol. ii., p. 256.

———— On the Copper-bearing Rocks of Alderley Edge, Cheshire. *Geol. Mag.*, vol. i., p. 65.

*Hull, E. and A. H. Green.* On the Millstone Grit of North Staffordshire and the adjoining parts of Derbyshire, Cheshire, and Lancashire. *Quart. Jour. Geol. Soc.*, vol. xx., p. 242.

*Sainter, J. D.* Sandstone Hammer in a Diluvial Deposit at Macclesfield. *Geologist*, vol. vii., p. 56.

*Taylor, J.* On the Drift Deposits in the Neighbourhood of Crewe, Cheshire. *Trans. Manchester Geol. Soc.*, vol. iv., p. 308.

1865.

*Binney, E. W.* A few Remarks on Mr. Hull's Additional Observations on the Drift deposits in the Neighbourhood of Manchester. *Mem. Lit. Phil. Soc. Manchester,* Ser. 3, vol. ii., p. 462.

*Boult. J.* On the Alleged Submarine Forests on the Shores of Liverpool Bay and the River Mersey. *Journ. Polytech. Soc. Liverpool.*

*Brodie. Rev. P. B.* Remarks on three outliers of Lias in North Shropshire and South Cheshire. Staffordshire, and Cumberland, and their correlation with the main range. *Proc. Warwick Field Club,* p. 6.

*Darbishire. R. D.* On the Genuineness of certain Fossils from the Macclesfield Drift-beds. *Geol. Mag.,* vol. ii., p. 293.

*Hull. E.* Additional Observations on the Drift deposits, and more recent Gravels in the Neighbourhood of Manchester. *Mem. Lit. Phil. Soc. Manchester.* Ser. 3, vol. ii., p. 449.

*Morton, G. H.* Section of the Strata from Hilbre to Huyton. Scale 2 inches to a mile.

———— On the Recent Shell-bed at Wallasey. *Proc. Liverpool Geol. Soc.* Session 6, p. 26.

*Smith. H. E.* A Record of Archæological Products of the Sea Shore of Cheshire in 1864 *Quart. Archæol. Journ. and Rev.*


1866.

*Boult. J.* Further Observations on the Alleged Submarine Forests on the Shores of Liverpool Bay and the River Mersey. In reply to Dr. Hume's Communication of July 10, 1865. *Trans. Hist. Soc. Lancash. and Chesh.,* N. Ser. vol. vi. p. 59.

*Hardwick. C.* A few Thoughts on Geology in its Relation to Archæology. *Trans. Manchester Geol. Soc.* vol. v., p. 201.

*Hume. Rev. Dr.* On the Changes in the Sea Coast of Lancashire and Cheshire. *Trans. Hist. Soc. Lancash. and Chesh.,* N. Ser., vol. vi. p. 1.

*Morton. G. H.* On the Geology of the Country bordering the Mersey and the Dee. *Proc. Liverpool Geol. Soc.* Session 7, p. 87. and *Liverpool Naturalists Journ.* No. 1, p 75.

*Ricketts. Dr. C.* On a Wooden Implement found in Bidston Moss. *Proc. Liverpool Geol. Soc.* Session 7, p. 8.

*Sainter. J. D.* Macclesfield Drift Shells, &c. *Trans. Manchester Geol. Soc.* vol. v. p 114  See also *Geol. Mag.* vol. ii, p 365

*Smith. H. E.* Notabilia of the Archæology and Natural History of the Mersey District during Three Years. 1863, 4, 5. pp. 306, 394. and Plate refer to section of beach. *Trans. Hist. Soc. Lancash. Chesh.* N. Ser. vol. vi., p 195


1867.

*Binney. E. W.* The Drift of the Western and Eastern Counties. *Geol. Mag.* vol. iv, p. 291.

*Morton. G. H.* On the Presence of Glacial Ice in the Valley of the Mersey during the Post Pliocene Period. *Proc. Liverpool Geol. Soc.* Session 8, p 4

*Smith. H. E.* Archæology of the Mersey District, 1866 *Trans. Hist. Soc. Lancash. Chesh.* N. Ser. vol.' vii., p. 169  Growth of trees in situ. pp 191-194

*Thomas. J. E.* Prize Essay upon the Encroachment of the Sea between the River Mersey and the Bristol Channel. 8vo. Lond.

*Williamson. Prof. W. C.* On a Cheirotherian Footprint from the Base of the Keuper Sandstone of Daresbury. Cheshire. *Quart. Journ. Geol. Soc.* vol. xxii., p 56


1868.

*Morton. G.* On the Disposition of Ice as represented in the Cheshire. &c. 8vo. Rev. *Quart. Journ. Geol. Soc.* vol. xxv. p. 351

1869.

Bonnel, R. The New Red Sandstone as a Source of Water Supply. Proc. Liverpool Geol. Soc., Session 10, p. 58.

Hull, E. On the Evidences of a Ridge of Lower Carboniferous Rocks crossing the Plain of Cheshire beneath the Trias, and forming the boundary between the Permian Rocks of the Lancashire Type on the North, and those of the Salopian Type on the South. Quart. Journ. Geol. Soc., vol. xxv. p. 171.

Picton, C. Observations on the Cheshire Coast. Proc. Liverpool Geol. Soc., Session 19, p. 37.

Scouler, Dr. J. D. The Geology and Archæology of some of the Macclesfield Drift Beds. Geol. and Nat. Hist. Repertory, vol. ii. p. 360.

1870.

Bonnel, R. The Mersey and the Dee—Their former Channels and Change of Level. Proc. Liverpool Geol. Soc., Session 11, p. 41.

De Rance, C. E. Notes on the Geology of the Country around Liverpool. Nature, vol. ii. No. 46, p. 587.

Eyton, Miss C. On the Age and Geological Position of the Bone Clay of the Western Counties. Geol. Mag., vol. vii. p. 543.

Maw, G. On the Occurrence of the Rhætic Beds in North Shropshire and Cheshire. Ibid., p. 301.

Taylor, J. E. Note on the Middle Drift-beds in Cheshire. Ibid., p. 142.

Wood, S. T., Jun. Observations on the Sequence of the Glacial Beds. 2nd part. Ibid., p. 63.

1871.

Report of the Commissioners appointed to inquire into the several matters relating to Coal in the United Kingdom. Vol. i. General Report, &c. J. Dickinson. Lancashire, Cheshire, &c., p. 16. Prof. A. C. Ramsay and others. Report of Committee D. appointed to inquire into the probability of Finding Coal under the Permian New Red Sandstone, &c., p. 155. ———— Vol. ii. General Minutes and Proceedings of Committees. J. Knowles. Temperature of Air and Soils, p. 192. On the Probability of Finding Coal under the Permian, &c., p. 413. F. S. Lead.

Aitken, J. The President's Address. Note on a Section at Stockport. Trans. Manchester Geol. Soc., vol. x. No. 1, p. 30.

———— On Faults in Drift at Stockport, Cheshire. Ibid., p. 48, and Geol. Mag., vol. viii. p. 117.

Binney, E. W. Notes on some of the High Level Drifts in the Counties of Chester, Derby, and Lancashire. Proc. Lit. Phil. Soc., Manchester, vol. x. p. 66.

———— On the Permian Strata of East Cheshire. Mem. Lit. Phil. Soc., Manchester, ser. 3, vol. 4. p. 214.

Buck, J. Speculations on the Former Topography of Liverpool and its Neighbourhood. Proc. Lit. Phil. Soc., Liverpool, No. xxv. p. 11. (Refers to changes of Herle Bank and Wallasey Pool, pp. 13–21.)

De Rance, C. E. On the Glacial Phenomena of Western Lancashire and Cheshire. Quart. Journ. Geol. Soc., vol. xxvi. p. 641.

———— On the Postglacial Deposits of Western Lancashire and Cheshire. Ibid., p. 653. (Both these papers published with the first number of vol. xxvii., and therefore in the year after the greater part of vol. xxvi.)

———— On the Pre-Glacial Geography of Northern Cheshire. Geol. Mag., vol. vii. p. 150.

Morton, G. H. On a continued Comparison of the Triassic Rocks around Liverpool. Rep. Brit. Assoc. for 1870. Trans. of Sections, p. 81.

———— Anniversary Address. Proc. Liverpool Geol. Soc., Session 12, p. 5.

1872.

Irton, M. S. Land-slip near Northwich, Cheshire. Science Gossip, No. 89, p. 116.

Mackintosh, D. On a Sea-coast Section of Boulder-clay in Cheshire. Quart. Journ. Geol. Soc., vol. xxviii. p. 382.

*Morton, G. H.* Minerals that occur in the Neighbourhood of Liverpool, with the Localities, &c. *Proc. Liverpool Geol. Soc.,* Session 13, p. 91.

———— Shells found in the Glacial Deposits around Liverpool, with the Localities. &c. *Ibid.,* p. 92.

*Morton, Dr.* [T.] On Geological Systems and Endemic Disease. *Rep. Brit. Assoc. for* 1871, *Trans. of Sections,* p. 107.

*Reade, T. M.* The Geology and Physics of the Post-Glacial Period, as shewn in the Deposits and Organic Remains in Lancashire and Cheshire. *Proc. Liverpool Geol. Soc.,* Session 13, p. 36.

———— The Post-Glacial Geology and Physiography of West Lancashire and the Mersey Estuary. *Geol. Mag.,* vol. ix., p. 111.

———— (Letter on boring for Coal.) *Liverpool Daily Post,* Sept. 16. (Noticed in *Nature,* No. 151, p. 421.)

*Ricketts, Dr. C.* Valleys, Deltas, Bays, and Estuaries. (President's Address.) *Proc. Liverpool Geol. Soc.,* Session 1871-2.

*Ward, T.* The Landslips at Northwich. *Nature,* vol. v., No. 119, p. 289.

### 1873.

*Anon.* (P. S.) Ancient Cheshire Forest [W. of Warrington]. *Science Gossip,* No. 99, pp. 67, 68.

*Boult J.* The Mersey as known to the Romans. *Proc. Lit. Phil. Soc. Liverpool,* No. xxvii., p. 249.

*De Rance, C. E.* The Cyclas clay of West Lancashire. *Geol. Mag.,* vol. x., p. 187.

———— The Lower Scrobicularia and Lower Cyclas Clays of the Mersey and the Ribble. *Ibid.,* p. 287.

*Dickinson, J.* Copy of a Report on the subject of Landslips in the Salt Districts, made to Her Majesty's Secretary of State for the Home Department. Fol. *London.*

*Mackintosh, D.* Observations on the more remarkable Boulders of the North-west of England and the Welsh Borders. *Quart. Journ. Geol. Soc.,* vol. xxix., p. 351.

*Morton, G. H.* The Strata below the Trias in the Country around Liverpool; and the Probability of Coal occurring at a Moderate Depth. *Proc. Lit. Phil. Soc. Liverpool,* No. xxvii, p. 157.

*Reade, T. M.* Formby and Leasowe Marine Beds, or the so-called " Cyclas Clay." *Geol. Mag.,* vol. x., p. 238.

———— The Buried Valley of the Mersey. *Proc. Liverpool Geol. Soc.,* Session 14, p. 42.

*Ward, T.* The Cheshire Salt District. *Proc. Lit. Phil. Soc. Liverpool,* No. xxvii., p. 39.

### 1874.

*Dickinson, J.* On the Saliferous Strata. *Trans. Geol. Soc. Manchester,* vol. xiii., pt. ii., p. 23.

*Moffat, Dr. T.* On Geological Systems and Endemic Diseases. *Rep. Brit. Assoc.* for 1873, Sections, p. 84.

*Morris, Prof. J.* Landslips and Sinkings in Cheshire. *Geol. Mag.,* Dec. ii., vol. i., p. 259.

*Reade, T. M.* The Drift beds of the North-West of England. *Quart. Jour. Geol. Soc.,* vol. xxx., p. 27.

*Ricketts, Dr. C.* Is the Mersey filling up? *Liverpool Daily Courier,* May 15.

*Shone, W.* Discovery of Foraminifera, &c., in the Boulder-Clays of Cheshire. *Quart. Jour. Geol. Soc.,* vol. xxx., p. 131.

### 1875.

*De Rance, C. E.* On the Relative Age of some Valleys in the North and South of England, and of the various [Glacial] and Post-Glacial Deposits occurring in them. *Proc. Geol. Assoc.,* vol. iv., No. 4, p. 221.

*Roberts, Isaac.* President's Address. *Proc. Liverpool Geol. Soc.,* vol. iii., pt. 1, p. 3.

*Sainter, J. D.* The Geology of Mow Cop, Congleton Edge, and the surrounding District. *N. Staff. Field Club Papers,* p. 140.

1876.

*Burghardt, [Dr.] C. S.* On the formation of Azurite from Malachite. *Proc. Lit. Phil. Soc. Manch.*, vol. xv., pp. 72, 73.

*Dawkins, Prof. W. B.* On the Water Supply in the Red Rocks of Lancashire, of Cheshire. *Trans. Manchester Geol. Soc.*, vol. xiv., pt. vi., p. 133.

*De Rance, C. T.* First Report of the Committee for investigating the circulation of the Underground Waters in New Red Sandstone and Permian Formations of England . . . . . *Rep. Brit. Assoc. for* 1875, p. 114.

*Laurance, J.* Some Impressions of the Feet of Animals recently found in the New Red Sandstone, near Liverpool. *Trans. Leicester Lit. Phil. Soc.*, pt. ii., p. 41.

*Mackintosh, D.* Result of Observations on the Eskers, Lake-basins, and Post-Glacial River-courses of Cheshire . . . . (*Chester Soc. Nat. Sci.*) Abstract in *Geol. Mag.*, Dec. ii., vol. iii., p. 272.

*Potter, C.* Observations on the Geology and Archæology of the Cheshire Shore. *Trans. Hist. Soc. Lanc. Chesh.*, Ser. 3, vol. iv., p. 121, pl. iii.

———— The So-called Forest Beds. 23*rd Ann. Rep. Brighton Nat. Hist. Soc.*, p. 10.

*Shoolbred, J. N.* On the Changes in the Tidal Portion of the River Medway and its Estuary. *Proc. Inst. Civ. Eng.*, vol. xlvi., p. 3, pls. 3 7. Discussion, p. 50.

1877.

*De Rance, C. E.* Second Report of the Committee for investigating the circulation of Underground Waters in New Red Sandstone and Permian Formations of England . . . . *Rep. Brit. Assoc. for* 1876, p. 95.

*Mackintosh, D.* On a Number of New Sections around the Estuary of the Dee . . . *Quart. Journ. Geol. Soc.*, vol. xxxiii., p. 730.

*Morton, G. H.* The Glacial Striæ of the Country around Liverpool. *Proc. Liverpool Geol. Soc.*, vol. iii., pt. 3, p. 284.

*Ricketts, Dr. C.* The Conditions existing during the Glacial Period; with an Account of the Glacial Deposits in the Valley between Tranmere and Oxton. *Proc. Liverpool Geol. Soc.*, vol. iii. pt. 3, p. 245.

*Williams, W. M.* Hog-Wallows and Prairie Mounds. *Nature*, vol. xvi. pp. 6, 7.

1878.

*Blower, B.* The Mersey, Ancient and Modern. 8vo. *Liverpool.* Pp. vii., 88.

*Crosskey, Rev. H. W.* Fifth Report of the Committee . . . . . for . . . . . recording the position (&c.) . . . . . of the Erratic Blocks of England. . . . . *Rep. Brit. Assoc. for* 1877, p. 81.

*Dawkins, Prof. W. B.* The Mammoth at Northwich. *Trans. Manchester Geol. Soc.*, vol. xv., pt. iii., p. 55.

*De Rance, C. E.* Third Report of the Committee for investigating the circulation of the Underground Waters in New Red Sandstone and Permian Formations of England. . . . . *Rep. Brit. Assoc. for* 1877, p. 56.

*Plant, John.* [The Geology in] Report of Excursion of Manchester Scientific Students Assoc. to Weston Point Quarries, Runcorn. *Manchester City News*, Aug. 3.

*Roberts, I.* Boring on East Hoyle Bank. *Proc. Lit. Phil. Soc. Liverpool*, No. xxxii., p. lxxxviii.

*Shone, W.* On the Glacial Deposits of West Cheshire, together with Lists of the Fauna found in the Drift of Cheshire. . . . . *Quart. Journ. Geol. Soc.*, vol. xxxiv., p. 383.

———— How we found the Microzoa in the Boulder Clays of Cheshire, &c. . . . . *Midland Naturalist*, vol. i., p. 292.

1879.

*Dawkins, Prof. W. B.* On the Range of the Mammoth in Space and Time. *Quart. Journ. Geol. Soc.*, vol. xxxv., p. 138.

*De Rance, C. E.* Fourth Report of the Committee for investigating the circulation of the Underground Waters in New Red Sandstone and Permian Formations of England. . . . . *Rep. Brit. Assoc. for* 1878, p. 382.

———— Fifth ditto. *Rep. Brit. Assoc. for* 1879, p. 155.

*Mackintosh, D.* Results of a Systematic Survey . . . . of the Directions and Limits of the Dispersion . . . . . of the Erratic Blocks . . . . . of the West of England . . . . . *Quart. Journ. Geol. Soc.*, vol. xxxv., p. 425.

*Ricketts, Dr. C.* On some remarkable Pebbles in the Boulder Clay. *Proc. Liverpool Geol. Soc.*, vol. iv., pt. 1, p. 10. Abstract in *Rep. Brit. Assoc.* for 1879, p. 39, plate (adding " of Cheshire and Lancashire " to title).

1880.

*Everett, Prof.* Thirteenth Report of the Committee . . . . . for . . . . . investigating the Rate of Increase of Underground Temperature. (Dukinfield.)   *Rep. Brit. Assoc.* for 1880, p. 26.

*Mackintosh, D.* On the Correlation of the Drift-deposits of the North-west of England with those of the Midland and Eastern Counties. *Quart. Journ. Geol. Soc.*, vol. xxxvi., p. 178.

*Reade, T. M.* A Problem for Irish Geologists in Post-Glacial Geology. *Journ. Roy. Geol. Soc. Ireland*, N. Ser., vol. v. (pt. iii.), p. 173.

*Spratt, Rear-Admiral T.* A Suggestion for the Improvement of the Entrance to the Mersey. Pp. 20, map. 8vo. London. (Origin of the sand-banks, &c., p. 9.)

1881.

*Shipman, J.* Notes on the Triassic Rocks of Cheshire and their Equivalents at Nottingham. *Nottingham Nat. Hist. Soc.* Report for 1880.

# INDEX.

www.ingramcontent.com/pod-product-compliance
Lightning Source LLC
Chambersburg PA
CBHW031747090426
42739CB00008B/912